This book provides a comprehensive look at the Schwarz–Christoffel transformation, including its history and foundations, practical computation, common and less common variations, and many applications in fields such as electromagnetism, fluid flow, design and inverse problems, and the solution of linear systems of equations. As the only modern treatment of this subject, it is an accessible resource for engineers, scientists, and applied mathematicians who seek more experience with theoretical or computational conformal mapping techniques. The most important theoretical results are stated and proved, but the emphasis throughout remains on concrete understanding and implementation, as evidenced by the 76 figures based on quantitatively correct illustrative examples. There are more than 150 classical and modern reference works cited for readers needing more details. There is also a brief appendix illustrating the use of the Schwarz–Christoffel Toolbox for MATLAB, the state-of-the-art package for computing these maps.

Tobin A. Driscoll is Assistant Professor of Mathematics at the University of Delaware. Professor Driscoll is the author of numerous articles that have appeared in such publications as *Science*, *SIAM Review*, *Journal of Computational Physics*, and *Physics of Fluids A*. He has received a SIAM Outstanding Paper Prize and won Second Prize in the Leslie Fox numerical analysis competition.

Lloyd N. Trefethen is Professor of Numerical Analysis and Head of the Numerical Analysis Group in the Oxford University Computing Laboratory. He is the author of *Numerical Linear Algebra* (1997, with David Bau III) and *Spectral Methods in MATLAB* (2000). He was the first winner of the Fox Prize in Numerical Analysis and is a frequent invited speaker at international conferences, including the most recent quadrennial International Congress of Mathematicians (Berlin, 1998). He has held professorial positions at New York University (Courant Institute), MIT, Cornell University, and Oxford University.

CAMBRIDGE MONOGRAPHS ON
APPLIED AND COMPUTATIONAL
MATHEMATICS

Series Editors
P. G. CIARLET, A. ISERLES, R. V. KOHN, M. H. WRIGHT

8   Schwarz–Christoffel Mapping

The Cambridge Monographs on Applied and Computational Mathematics reflect the crucial role of mathematical and computational techniques in contemporary science. The series publishes expositions on all aspects of applicable and numerical mathematics, with an emphasis on new developments in this fast-moving area of research.

State-of-the-art methods and algorithms as well as modern mathematical descriptions of physical and mechanical ideas are presented in a manner suited to graduate research students and professionals alike. Sound pedagogical presentation is a prerequisite. It is intended that books in the series will serve to inform a new generation of researchers.

Also in this series:

A Practical Guide to Pseudospectral Methods, *Bengt Fornberg*
Dynamical Systems and Numerical Analysis, *A. M. Stuart and A. R. Humphries*
Level Set Methods, *J. A. Sethian*
The Numerical Solution of Integral Equations of the Second Kind,
    Kendall E. Atkinson
Orthogonal Rational Functions, *Adhemar Bultheel, Pablo González-Vera,*
    Erik Hendriksen, and Olav Njåstad
Theory of Composites, *Graeme W. Milton*
Geometry and Topology for Mesh Generation, *Herbert Edelsbrunner*

Schwarz–Christoffel Mapping

Tobin A. Driscoll
University of Delaware

Lloyd N. Trefethen
Oxford University

CAMBRIDGE
UNIVERSITY PRESS

CAMBRIDGE UNIVERSITY PRESS
Cambridge, New York, Melbourne, Madrid, Cape Town, Singapore,
São Paulo, Delhi, Dubai, Tokyo, Mexico City

Cambridge University Press
The Edinburgh Building, Cambridge CB2 8RU, UK

Published in the United States of America by Cambridge University Press, New York

www.cambridge.org
Information on this title: www.cambridge.org/9780521807265

First published 2002

A catalogue record for this publication is available from the British Library

Library of Congress Cataloguing in Publication Data
Driscoll, Tobin A. (Tobin Allen), 1969–
Schwarz–Christoffel mapping / Tobin A. Driscoll, Lloyd N. Trefethen.
p.   cm.—(Cambridge monographs on applied and computational mathematics ; v. 8)
ISBN 0-521-80726-3
1. Conformal mapping.   I. Trefethen, Lloyd N. (Lloyd Nicholas)   II. Title.
III. Cambridge monographs on applied and computational mathematics ; 8.
QA360 .D75   2002
516.3'6–dc21   2001043099

ISBN 978-0-521-80726-5 Hardback

To Luke, Emma, and Jacob.

Every valley shall be exalted, and every mountain and hill shall be made low: and the crooked shall be made straight, and the rough places plain.

Isaiah 40:4

Contents

Figures

Preface

In the autumn of 1978, Peter Henrici took leave from the ETH in Zurich to visit the Numerical Analysis Group at Stanford University. The second author, then a graduate student, asked Henrici if he might propose a project in the area of computational complex analysis. Henrici's suggestion was, why don't you see what you can do with the Schwarz–Christoffel transformation?

For months thereafter LNT spent all of every weekend working on SC mapping at the computer terminals of the Stanford Linear Accelerator Center. This brief but intense project led to one of the first technical reports ever printed in TEX, which was published in the first issue of the *SIAM Journal on Scientific and Statistical Computing*; to the FORTRAN package SCPACK; and to a lasting love of numerical conformal mapping. In the following years it led further to extensions and applications of SC ideas carried out in collaboration with various people, including Alan Elcrat and Frédéric Dias on free-streamline flows, Ruth Williams on oblique derivative problems, and Louis Howell on modified formulas for elongated regions.

By the early 1990s, LNT was a faculty member at Cornell University and the first author was a graduate student. We worked together on a number of topics from hydrodynamic stability to "Can one hear the shape of a drum?" but the subject we kept coming back to was Schwarz–Christoffel mapping. Once again it started with a brief suggestion. MATLAB graphics had become widespread, and LNT proposed to TAD, why not convert SCPACK to an interactive MATLAB package? The resulting SC Toolbox was even more fun than we had imagined. We quickly became accustomed to drawing a polygon with a mouse, clicking a button, and seeing in seconds a plot of a conformal map correct to eight digits. The SC Toolbox was one of the earliest advanced applications built around a MATLAB graphical user interface. It matured as MATLAB did, quickly eclipsing SCPACK by surpassing it not just in convenience but also in capabilities and numerical power. (The second author insists on adding that a parallel development occurred as TAD quickly overtook LNT in his knowledge of Schwarz–Christoffel mapping.)

Mathematically, the subject continued to grow. Others such as Henrici (before his death in 1986), Däppen, and Floryan and Zemach had made important contributions. A new collaborator was Steve Vavasis, who together with TAD developed the CRDT algorithm. Based on Delaunay triangulation and Möbius transformations, it is very different from other methods in concept and largely immune to the problems of exponential crowding. Applications to large-scale nonsymmetric matrix iterations were pursued in collaboration with Kim-Chuan Toh. Applications to the computation of Green's functions and their uses in approximation theory and digital filtering were developed with Mark Embree. We came to realize that although experts in complex analysis had a depth of theoretical understanding that we could not match, we had accumulated an experience of the practical side of SC problems, and of the range of their variations and applications, that was unique.

The result is this book. Our style is concise, covering more material than might be expected in the small number of pages. For nearly every topic we have included a number of computer-generated figures, all of which are not just schematic but quantitatively correct. We hope that these figures will quickly tell the reader *what* one can do with these methods—always with high speed and accuracy—and that careful reading if desired will reveal the details of *how*. To assist in this latter purpose, an appendix gives short MATLAB scripts showing how to generate some of our figures with the help of the SC Toolbox.

We thank those whose comments and contributions improved this book. Endre Süli and Markus Melenk helped us link the one-half quadrature rule with mesh refinement in finite elements; Lehel Banjai conducted experiments regarding the refinement ratio. Tom DeLillo, Alan Elcrat, and John Pfaltzgraff shared an advance copy of their work on doubly connected maps and thereby sparked our discussion in section 4.9. Louis Howell generously created the data for the circular-arc mappings in Figure 4.24. Martin Gutknecht and Dieter Gaier shared many discussions of conformal mapping with the second author over the years. For inspiration, we are grateful to Peter Henrici, who in many ways is the person most responsible for this book.

1

Introduction

1.1 The Schwarz–Christoffel idea

The idea behind the Schwarz–Christoffel (SC) transformation and its variations is that a conformal transformation f may have a derivative that can be expressed as

$$f' = \prod f_k \tag{1.1}$$

for certain canonical functions f_k. A surprising variety of conformal maps can be fitted into this basic framework. In fact, virtually all conformal transformations whose analytic forms are known are Schwarz–Christoffel maps, albeit sometimes disguised by an additional change of variables.

Geometrically speaking, the significance of (1.1) is that

$$\arg f' = \sum \arg f_k.$$

In the classical transformation, each $\arg f_k$ is designed to be a step function, so the resulting $\arg f'$ is piecewise constant with specific jumps (i.e., f maps the real axis onto a polygon). To be specific, let P be the region in the complex plane $\mathbf{C}$ bounded by a polygon Γ with vertices $w_1, \ldots, w_n$, given in counterclockwise order, and interior angles $\alpha_1\pi, \ldots, \alpha_n\pi$. For now, we assume that P is bounded and without cusps or slits, so that $\alpha_k \in (0, 2)$ for each k. Let f be a conformal map of the upper half-plane H^+ onto P, and let $z_k = f^{-1}(w_k)$ be the kth **prevertex**.[1] We shall assume $z_n = \infty$ without loss of generality, for if infinity is not already a prevertex, we can simply introduce its image (which lies

---

[1] The Carathéodory–Osgood theorem [Hen74] guarantees a continuous extension of f to the boundary. Hence the prevertices are well defined.

1

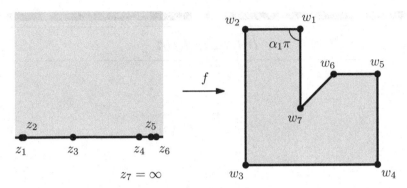

Figure 1.1. Notational conventions for the Schwarz–Christoffel transformation. In this case, z_1 and z_2 are mathematically distinct but graphically difficult to distinguish. As with all figures in this book, everything shown is not just schematic but also quantitatively correct.

on Γ) as a new vertex with interior angle π. The other prevertices $z_1, \ldots, z_{n-1}$ are real. Figure 1.1 illustrates these definitions.

As with all conformal maps, the main effort is in getting the boundary right. By the Schwarz reflection principle, which was invented for this purpose, f can be analytically continued across the segment (z_k, z_{k+1}). In particular, f' exists on this segment, and we see that $\arg f'$ must be constant there. Furthermore, $\arg f'$ must undergo a specific jump at $z = z_k$, namely

$$\left[\arg f'(z)\right]_{z_k^-}^{z_k^+} = (1 - \alpha_k)\pi = \beta_k \pi. \tag{1.2}$$

The angle $\beta_k \pi$ is the **turning angle** at vertex k. We now identify a function f_k that is analytic in H^+, satisfies (1.2), and otherwise has $\arg f_k$ constant on $\mathbf{R}$:

$$f_k = (z - z_k)^{-\beta_k}. \tag{1.3}$$

Any branch consistent with H^+ will work; to be definite, we pick the branch with $f_k(z) > 0$ if $z > z_k$ on $\mathbf{R}$. The action of f_k on the real line is sketched in Figure 1.2.

The preceding argument suggests the form

$$f'(z) = C \prod_{k=1}^{n-1} f_k(z)$$

for some constant C. We will prove the following fundamental theorem of Schwarz–Christoffel mapping in section 2.2.

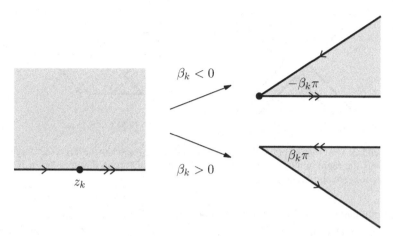

Figure 1.2. Action of a term (1.3) in the SC product. In either case, the argument of the image jumps by $\beta_k \pi$ at z_k.

Theorem 1.1. *Let P be the interior of a polygon Γ having vertices $w_1, \ldots, w_n$ and interior angles $\alpha_1 \pi, \ldots, \alpha_n \pi$ in counterclockwise order. Let f be any conformal map from the upper half-plane H^+ to P with $f(\infty) = w_n$. Then*

$$f(z) = A + C \int^z \prod_{k=1}^{n-1} (\zeta - z_k)^{\alpha_k - 1} \, d\zeta \tag{1.4}$$

for some complex constants A and C, where $w_k = f(z_k)$ for $k = 1, \ldots, n - 1$.

The lower integration limit is left unspecified, as it affects only the value of A.

The formula also applies to polygons that have slits ($\alpha = 2$) or vertices at infinity ($-2 \leq \alpha \leq 0$). Indeed, arbitrary real exponents can meaningfully appear in (1.4), although the resulting region may overlap itself and not be bounded by a polygon in the usual sense of the term; see section 4.7.

Formula (1.4) can be adapted to maps from different regions (such as the unit disk), to exterior maps, to maps with branch points, to doubly connected regions, to regions bounded by circular arcs, and even to piecewise analytic boundaries. These and other variations are the subject of Chapter 4.

But there is a major difficulty we have not yet mentioned: without knowledge of the prevertices z_k, we cannot use (1.4) to compute values of the map. In view of how we arrived at (1.4), the image $f(\mathbf{R} \cup \{\infty\})$ of the extended real line will necessarily be *some* polygon whose interior angles match those of P, no matter what real values of z_k are used; that much is forced by the parameters α_k. (Here we are broadening the usual idea of "polygon" to allow

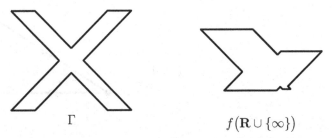

Γ $f(\mathbf{R} \cup \{\infty\})$

Figure 1.3. The effect of prevertices on side lengths. The region on the left is the "target," whereas the region on the right illustrates the type of distortion that may occur if the prevertices are chosen incorrectly.

self-intersections.) The prevertices, however, influence the side lengths of $f(\mathbf{R} \cup \{\infty\})$, as illustrated in Figure 1.3. Determining their correct values for a given polygon is the Schwarz–Christoffel **parameter problem**, and its solution is the first step in using the SC formula.[2] In sections 2.3–2.5 we will consider some of the classical cases for which the parameter problem can be solved explicitly.

In the majority of practical problems, there is no analytic solution for the prevertices, which depend nonlinearly on the side lengths of Γ. Numerical computation is also usually needed to evaluate the integral in (1.4) and to invert the map. Thus, much of the potential of SC mapping went unrealized until computers became readily available in the last quarter of the twentieth century. Numerical issues are discussed in Chapter 3.

1.2 History

The roots of conformal mapping lie early in the nineteenth century. Gauss considered such problems in the 1820s. The Riemann mapping theorem was first stated in Riemann's celebrated doctoral dissertation of 1851: any simply connected region in the complex plane can be conformally mapped onto any other, provided that neither is the entire plane.[3] The Schwarz–Christoffel formula was discovered soon afterwards, independently by Christoffel in 1867 and Schwarz in 1869.

---

[2] Sometimes the constants A and C are included as unknowns in the parameter problem. However, they can be found easily once the prevertices are known, for they just describe a scaling, rotation, and translation of the image.

[3] Riemann's proof, based on the Dirichlet principle, was later pointed out by Weierstrass to be incomplete. Rigorous proofs did not appear until the work of Koebe, Osgood, Carathéodory, and Hilbert early in the twentieth century.

Elwin Bruno Christoffel (1829–1900) was born in the German town of Montjoie (now Monschau) and was studying mathematics in Berlin under Dirichlet and others when Riemann's dissertation appeared.[4] Christoffel completed his doctoral degree in 1856 and in 1862 succeeded Dedekind as a professor of mathematics at the Swiss Federal Institute of Technology in Zurich. It was in Zurich that he published the first paper on the Schwarz–Christoffel formula, with the Italian title, "Sul problema delle temperature stazonarie e la rappresentazione di una data superficie" [Chr67]. Christoffel's motivation was the problem of heat conduction, which he approached by means of the Green's function. This paper presented the discovery that, in the case of a polygonal domain, the Green's function could be obtained via a conformal map from the half-plane, as in (1.4). In subsequent papers he extended these ideas to exteriors of polygons and to curved boundaries [Chr70a, Chr70b, Chr71].

Hermann Amandus Schwarz (1843–1921) grew up nearly a generation after Christoffel but also very much under the influence of Riemann. In the late 1860s he was living in Halle, where his discovery of the Schwarz–Christoffel formula apparently came independently of Christoffel's. His three papers on the subject [Sch69a, Sch69b, Sch90] cover much of the same territory as Christoffel's, including the generalizations to curved boundaries (section 4.11) and to circular polygons (section 4.10), but the emphasis is quite different—more numerical and more concerned with particular cases such as triangles in [Sch69b] and quadrilaterals in [Sch69a].[5] Schwarz even published the world's first plot of a Schwarz–Christoffel map, reproduced in Figure 1.4. Schwarz's papers included his famous reflection principle: if an analytic function f, extended continuously to a straight or circular boundary arc, maps the boundary arc to another straight or circular arc, then f can be analytically continued across the arc by reflection.

In 1869 Christoffel moved briefly to the Gewerbeakademie in Berlin, and Schwarz succeeded him in Zurich. By this time the two were well aware of each other's work; the phrase *Schwarz–Christoffel transformation* is now nearly universal (although the order of the names is reversed in some of the literature of the former Soviet Union).

In the 130 years since its discovery, the Schwarz–Christoffel formula has had an extensive impact in theoretical complex analysis, especially as a constructive

---

[4] For extensive biographical information on Christoffel, the reader is referred to the sesquicentennial volume [BF81], particularly Pfluger's paper therein on Christoffel's work on the SC formula.

[5] Schwarz also credits Weierstrass for proving the existence of a solution for the unknown parameters (which Schwarz proved for $n = 4$) in the general case.

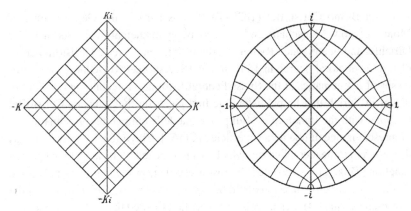

Figure 1.4. Schwarz's 1869 plot of the conformal map of a square onto a disk, reproduced from [Sch69b].

tool for proving the Riemann mapping theorem and related results. Its practical implementation—the main subject of this book—lagged far behind. Schwarz himself was the first to point out the importance of the parameter problem (discussed in the preceding section). This problem limited practical use to simple special cases, until the invention of computers.

Algorithmic discussions of the computation of Schwarz–Christoffel maps to prescribed polygons appear in several books, including those of Kantorovich and Krylov [KK64] and Gaier [Gai64]. Algorithms and in some cases computer programs have also appeared in numerous technical articles over the years, but in most of the earlier cases the authors were unaware of each other's work, and the quality of the result was wanting. Crucial issues that were often neglected included efficient evaluation of the SC integral and the need to impose necessary ordering conditions on the prevertices while solving the parameter problem. The most generally applicable computer programs for the classical problem are those of Trefethen [Tre80] (SCPACK) and Driscoll [Dri96] (SC Toolbox). The former was developed around 1980, and the latter began development in 1993. Both have been widely disseminated in the public domain.

Here is a list, more or less chronological, of contributors to constructive SC mapping of whom we are aware.

Gauss (1820s): Idea of conformal mapping
Riemann (1851): Riemann mapping theorem
Christoffel [Chr67, Chr70a, Chr70b, Chr71]: Discovery of SC formula and variants
Schwarz [Sch69a, Sch69b, Sch90]: Discovery of SC formula and variants
Kantorovich & Krylov [KK64] (first published 1936)

Polozkii (1955)
Filchakov ([Fil61], 1968, 1969, 1975)
Binns [Bin61, Bin62, Bin64]
Pisacane & Malvern [PM63]
Savenkov (1963, 1964)
Gaier [Gai64]: Book on numerical conformal mapping
Haeusler (1966)
Lawrenson & Gupta [LG68]: Adaptive quadrature, equations solver for parameters
Beigel (1969)
Hoffman (1971, 1974)
Gaier [Gai72]: "Crowding" phenomenon
Howe [How73]
Vecheslavov, Tolstobrova & Kokoulin [VT73, VK74]: Doubly connected regions
Foster & Anderson [FA74, And75]
Cherednichenko & Zhelankina [CZ75]
Squire [Squ75]
Meyer [Mey79]: Comparison of algorithms
Nicolaide [Nic77]
Prochazka [HP78, Pro83]: FORTRAN package KABBAV
Davis et al. [Dav79, ADHE82, SD85]: Curved boundaries
Hopkins & Roberts [HR79]: Solution by Kufarev's method
Reppe [Rep79]: First fully robust algorithm
Binns, Rees & Kahan [BRK79]
Volkov [Vol79, Vol87, Vol88]
Trefethen [Tre80, Tre84, Tre89, Tre93]: Robust algorithm, SCPACK, generalized parameter problems
Brown [Bro81]
Tozoni [Toz83]
Hoekstra (1983, [Hoe86]): Curved boundaries, doubly connected regions
Sridhar & Davis [SD85]: Strip maps
Floryan & Zemach [Flo85, Flo86, FZ87]: Channel flows, periodic regions
Bjørstad & Grosse [BG87]: Software for circular-arc polygons
Dias, Elcrat & Trefethen [ET86, DET87, DE92]: Free-streamline flows
Däppen [Däp87, Däp88]: Doubly connected regions
Costamagna ([Cos87, Cos01]): Applications in electricity and magnetism
Howell & Trefethen [How90, HT90, How93, How94]: Integration methods, elongated regions, circular-arc polygons
Pearce [Pea91]: Gearlike domains

Chaudhry [Cha92, CS92]: Piecewise smooth boundaries
Gutlyanskii & Zaidan [GZ94]: Kufarev's method
Driscoll [Dri96]: SC Toolbox for MATLAB
Hu [Hu95, Hu98]: Doubly connected regions (FORTRAN package DSCPACK)
Driscoll & Vavasis [DV98]: CRDT algorithm based on cross-ratios
Jamili (1999): Doubly connected regions

For more background information on conformal mapping in general and Schwarz–Christoffel mapping in particular, see [AF97, BF81, Hen74, Neh52, SL91, TD98, vS59, Wal64].

Essentials of Schwarz–Christoffel mapping

2.1 Polygons

For the rest of this book, a (generalized) **polygon** Γ is defined by a collection of vertices $w_1, \ldots, w_n$ and real interior angles $\alpha_1 \pi, \ldots, \alpha_n \pi$. It is convenient for indexing purposes to define $w_{n+1} = w_1$ and $w_0 = w_n$. The vertices, which lie in the extended complex plane $\mathbf{C} \cup \{\infty\}$, are given in counterclockwise order with respect to the interior of the polygon (i.e., locally the polygon is "to the left" as one traverses the side from w_k to w_{k+1}).

The interior angle at vertex k is defined as the angle swept from the outgoing side at w_k to the incoming side. If $|w_k| < \infty$, we have $\alpha_k \in (0, 2]$. If $\alpha_k = 2$, the sides incident on w_k are collinear, and w_k is the tip of a slit. The definition of the interior angle is applied on the Riemann sphere if $w_k = \infty$. In this case, $\alpha_k \in [-2, 0]$. See Figure 2.1. Specifying α_k is redundant if w_k and its neighbors are finite, but otherwise α_k is needed to determine the polygon uniquely.

In addition to the preceding restrictions on the angles α_k, we require that the polygon make a total turn of 2π. That is,

$$\sum_{k=1}^{n} (1 - \alpha_k) = 2, \tag{2.1}$$

or, equivalently,

$$\sum_{k=1}^{n} \alpha_k = n - 2.$$

We shall also, unless explicitly stated otherwise, require the polygon to be **simple** (forbid it from intersecting itself and thus covering part of the plane more than once). This condition has no elementary expression in terms of the

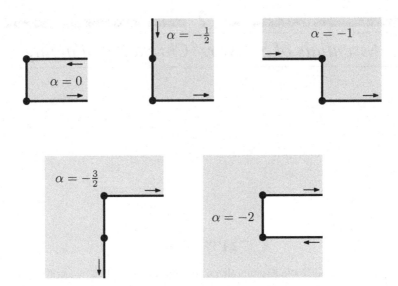

Figure 2.1. Examples of interior angles corresponding to a vertex at infinity.

vertices and angles—in a sense, it is artificial. We may occasionally use the term *polygon* to refer to a region bounded by a polygon. Context should keep the meaning clear.

2.2 The Schwarz–Christoffel formula

We now complete the proof of the half-plane formula of Theorem 1.1.

Theorem 2.1. *Let P be the interior of a polygon Γ having vertices $w_1, \ldots, w_n$ and interior angles $\alpha_1 \pi, \ldots, \alpha_n \pi$ in counterclockwise order. Let f be any conformal map from the upper half-plane H^+ to P with $f(\infty) = w_n$. Then*

Schwarz–Christoffel formula for a half-plane

$$f(z) = A + C \int^z \prod_{k=1}^{n-1} (\zeta - z_k)^{\alpha_k - 1} \, d\zeta \tag{2.2}$$

for some complex constants A and C, where $w_k = f(z_k)$ for $k = 1, \ldots, n-1$.

Proof. For simplicity, we treat just the case where all prevertices are finite and the product ranges over indices 1 to n. By the Schwarz reflection principle, the mapping function f can be analytically continued into the lower half-plane; the image continues into the reflection of P about one of the sides of Γ. By reflecting again about a side of the new polygon, we can return analytically to the upper half-plane. The same can be done for any even number of reflections of P, each time creating a new branch of f. The image of each branch must be a translated and rotated copy of P. Now, if A and C are any complex constants, then

$$\frac{(A + Cf(z))''}{(A + Cf(z))'} = \frac{f''(z)}{f'(z)}.$$

Therefore, the function f''/f' can be defined by continuation as a single-valued analytic function everywhere in the closure of H^+, except at the prevertices of Γ (where derivatives may fail to exist). Similarly, considering odd numbers of reflections, we see that f''/f' is single-valued and analytic in the lower half-plane as well.

We argued in the introduction that at a prevertex z_k,

$$f'(z) = (z - z_k)^{\alpha_k - 1} \psi(z)$$

for a function $\psi(z)$ analytic in a neighborhood of z_k. Therefore, f''/f' has a simple pole at z_k with residue $\alpha_k - 1$, and

$$\frac{f''(z)}{f'(z)} - \sum_{k=1}^{n} \frac{\alpha_k - 1}{z - z_k} \tag{2.3}$$

is an entire function. Because all the prevertices are finite, f is analytic at $z = \infty$, and a Laurent expansion there implies that $f''(z)/f'(z) \to 0$ as $z \to \infty$. By Liouville's theorem, it follows that the expression in (2.3) is identically zero. Expressing f''/f' as $(\log f')'$ and integrating twice results in (2.2). $\qquad\square$

An equally important version of the formula applies to the conformal map from the unit disk E.

Theorem 2.2. *Let P be the interior of a polygon Γ having vertices $w_1, \ldots, w_n$ and interior angles $\alpha_1 \pi, \ldots, \alpha_n \pi$ in counterclockwise order. Let f be any conformal map from the unit disk E to P. Then*

Schwarz–Christoffel formula for a disk

$$f(z) = A + C \int^{z} \prod_{k=1}^{n} \left(1 - \frac{\zeta}{z_k}\right)^{\alpha_k - 1} d\zeta \tag{2.4}$$

for some complex constants A and C, where $w_k = f(z_k)$ for $k = 1, \ldots, n$.

This variation can be derived from the basic SC principle; see section 4.1. The only substantive difference between (2.2) and (2.4) is that the product runs over all n prevertices in the latter case. The form of the integrand appears to be slightly different, but in fact it is a constant multiple of the original form. The reason for the change is so that the branch cuts of the integrand will point away from the origin if the principal branch of the logarithm is used to compute them. (Mistakes in branch cuts are a common pitfall in numerical conformal mapping, and writing a form that respects the "standard" cut is helpful in avoiding later trouble.)

2.3 Polygons with one or two vertices

As was pointed out in section 1.1, the SC formulas are only quasi-explicit. We must determine the prevertices z_k and affine constants A and C before we can compute point values of the map. There is some flexibility in the selection of these parameters. By the Riemann mapping theorem (or by considering Möbius self-maps of the upper half-plane), we can choose any three points on $\mathbf{R} \cup \{\infty\}$ to map to any three points of Γ, as long as their ordering is preserved. In other words, there are three degrees of freedom in the map, and that allows us to choose three prevertices arbitrarily. Hence if $n \leq 3$, there is no parameter problem to solve, and the SC formula becomes explicit.

The only polygon with $n = 1$ is a line, with vertex $w_1 = \infty$ and $\alpha_1 = -1$. Applying the half-plane formula (2.2) yields

$$f(z) = A + Cz, \tag{2.5}$$

which allows for scaling, rotation, and translation. The disk map formula (2.4) leads to

$$f(z) = A + C \int^z (\zeta - z_1)^{-2} d\zeta = A + \frac{C}{z - z_1}. \tag{2.6}$$

(The constant C is generic and automatically absorbs constant factors such as -1.) This confirms that any map from the disk to a half-plane is a Möbius transformation. There are two degrees of freedom still unspecified; one way to pin them down is to designate the image of zero, or the **conformal center**. Figure 2.2 shows an example.

If $n = 2$, then by (2.1) we have $\alpha_1 + \alpha_2 = 0$, so either $\alpha_1 = \alpha_2 = 0$ or $\alpha_1 = -\alpha_2 \neq 0$. In the first case, both vertices are at infinity and the region P is a strip. For the map from the half-plane H^+,

$$f(z) = A + C \int^z (\zeta - z_1)^{-1} d\zeta = A + C \log(z - z_1), \tag{2.7}$$

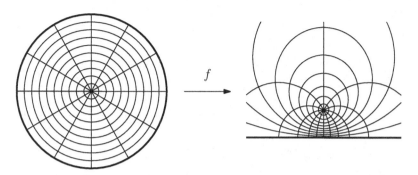

Figure 2.2. Map (2.6) from the disk to a half-plane ($n = 1$).

which is illustrated in Figure 2.3. The formula for the map from the disk E is

$$f(z) = A + C \int^z (\zeta - z_1)^{-1}(\zeta - z_2)^{-1} \, d\zeta$$

$$= A + C \int^z \left(\frac{1}{\zeta - z_1} - \frac{1}{\zeta - z_2} \right) d\zeta$$

$$= A + C \log \left(\frac{z - z_1}{z - z_2} \right), \tag{2.8}$$

and the result is shown in Figure 2.4.

By symmetry, the upper half of Figure 2.4 represents the map from the upper half-disk to a strip of half the original width. The map in the lower half-disk is simply a reflection about the real axis. If we reflect instead about the upper half-circle, the result, shown in Figure 2.5, is another sort of map from H^+ to the strip. The difference from Figure 2.3 is that both ends of the strip have finite preimages. (That equation (2.8) should describe the map from either H^+ or E is

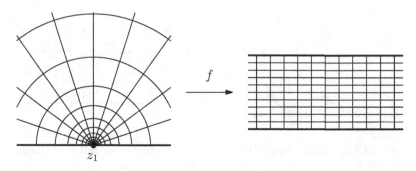

Figure 2.3. Map (2.7) from the upper half-plane to a strip ($n = 2$).

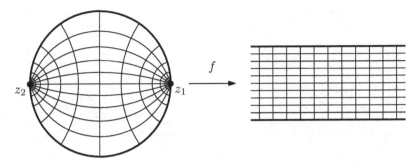

Figure 2.4. Map (2.8) from a disk to a strip with $z_1 = 1$, $z_2 = -1$ ($n = 2$).

not surprising, because the SC formulas are the same if $z_n \neq \infty$ in the half-plane case.) By further reflections we can obtain still more maps, such as one from the doubly slit plane shown in Figure 2.6. Subsets of this region, such as those shown in Figure 2.7, can also be mapped using just the disk formula (2.8). As we shall see in section 4.6 and elsewhere, reflection and subdivision are general tools that extend the applicability of Schwarz–Christoffel mapping.

In the case $n = 2$ with $\alpha_1 = -\alpha_2 \neq 0$, one vertex is finite and the other is infinite. Without loss of generality, suppose $w_2 = \infty$. The region P is a wedge whose vertex is w_1. The half-plane map defined by (2.2) is

$$f(z) = A + C \int^z (\zeta - z_1)^{\alpha_1 - 1} \, d\zeta$$
$$= A + C(z - z_1)^{\alpha_1}, \tag{2.9}$$

which is just a recovery of the fundamental SC building block. The map is shown in Figure 2.8 for two values of α_1. Observe the qualitatively different

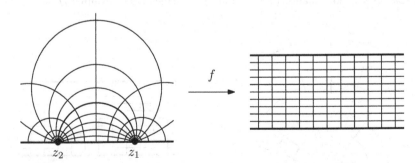

Figure 2.5. Use of (2.8) to map the upper half-plane to a strip whose ends are the images of -1 and 1.

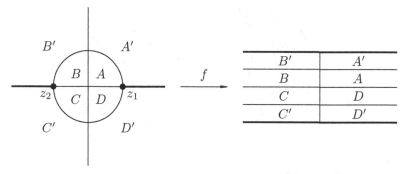

Figure 2.6. Reflection pattern for a map from a doubly slit plane to a strip using (2.8).

behavior for $\alpha_1 < 1$ and $\alpha_1 > 1$. The difference in these cases is in f', which is infinite or zero, respectively, at z_1. This difference causes the grid lines either to avoid the corner or to crowd there. The case $\alpha_1 < 1$ is sometimes called a **salient** or convex corner of P, while $\alpha_1 > 1$ is called a **reentrant** or concave corner. This distinction plays an important role in elliptic partial differential equations and their corresponding physical applications. For example, lightning tends to strike sharply pointed objects, and solids tend to fracture from cracks originating at sharp corners. The stress concentration suggested by the reentrant case in Figure 2.8 explains why ship portholes and airplane windows are designed without sharp corners.

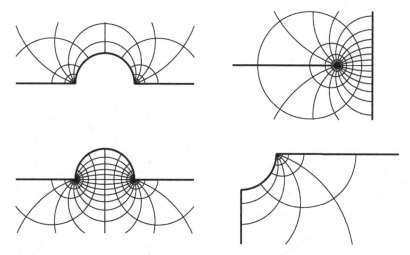

Figure 2.7. Other regions for which maps can be obtained by reflection using (2.8).

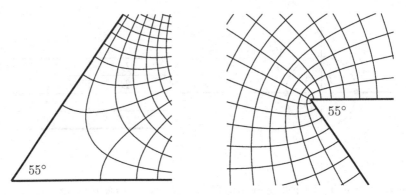

Figure 2.8. Maps from the half-plane to wedges of interior angles 55° and 305° ($n = 2$). In each case, the image of an equispaced cartesian grid is shown.

2.4 Triangles

Polygons with three vertices are the most general domains for which the Schwarz–Christoffel prevertices can be chosen arbitrarily, provided they remain distinct and properly ordered. Indeed, the mapping of triangles was the main focus of Schwarz's original paper [Sch69b] and was used to construct the map to the square shown in Figure 1.4. In this section, we use *triangle* to refer to unbounded polygons as well as the more conventional kind—in fact, the unbounded examples are the most interesting cases.

Because $\alpha_1 + \alpha_2 + \alpha_3 = 1$, at most two vertices may be infinite. If two vertices, say w_1 and w_3, are infinite, then $\alpha_2 \geq 1$ and the corner at w_2 is reentrant. The region P is therefore a strip with a kink in one side. A natural way to display such a map is to transplant a cartesian grid from an unkinked strip to the disk and thence to the kinked strip. Figure 2.9 shows two such maps. (Computation of

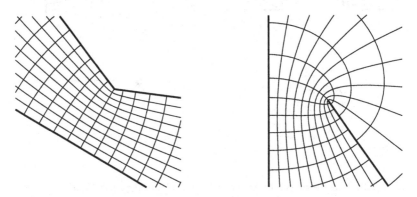

Figure 2.9. Maps to triangles with two infinite vertices ($n = 3$). The grids are conformal images of a regular grid in an infinite unkinked strip.

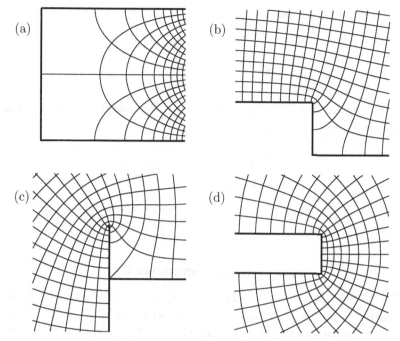

Figure 2.10. Maps from the half-plane to triangles with one infinite vertex. The curves are images of regular cartesian grids. The letters refer to formulas (2.10a)–(2.10d).

the map without using the intermediate disk, which is better numerical practice, is considered in section 4.2.) When $\alpha_2 = 2$, the kink folds back upon itself to form a slit. If additionally $\alpha_1 = \alpha_3 = -1/2$, the result is a slit half-plane region such as the one shown on the top right in Figure 2.7.

Triangles with one infinite vertex include some regions of practical and classical significance. In Figure 2.10 we show some of these maps. The formula for case (a) is elementary. If we let $w_3 = \infty$, $z_1 = -1$, and $z_2 = 1$, then

$$f(z) = A + C \int^z (\zeta^2 - 1)^{-1/2}\, d\zeta = A + C \cosh^{-1}(z). \qquad (2.10a)$$

Case (b), the map to a step, is only a little harder:

$$\begin{aligned} f(z) &= A + C \int^z \left(\frac{\zeta + 1}{\zeta - 1}\right)^{1/2} d\zeta \\ &= A + C \int^z \frac{\zeta + 1}{(\zeta^2 - 1)^{1/2}}\, d\zeta \\ &= A + C\left[(z^2 - 1)^{1/2} + \cosh^{-1}(z)\right]. \end{aligned} \qquad (2.10b)$$

Maps (c) and (d) can likewise be found in closed form:

$$f(z) = A + C(z - 1)^{1/2}(z + 5),\tag{2.10c}$$

$$f(z) = A + C\left[z(z^2 - 1)^{1/2} - \cosh^{-1}(z)\right].\tag{2.10d}$$

The SC map to any triangle can be compactly written using the incomplete beta function [GR94],

$$B_z(p, q) = \int_0^z \zeta^{p-1}(1 - \zeta)^{q-1}d\zeta.$$

If we choose $z_1 = 0$, $z_2 = 1$, then the map can be expressed as $f(z) = A + C B_z(\alpha_1, \alpha_2)$.

2.5 Rectangles and elliptic functions

If $n = 4$, the Schwarz–Christoffel prevertices are not ours to choose. In the general case there is no simple analytic determination of the one degree of freedom in the prevertices. However, in the important case when P is the interior of a rectangle, symmetry allows an explicit solution.

We rotate and translate the rectangle so that its vertices are $w_1 = -K + iK'$, $w_2 = -K$, $w_3 = K$, and $w_4 = K + iK'$. (The motivation for the notation will become clear shortly.) By symmetry, we choose the prevertices as $z_1 = -m^{-1/2}$, $z_2 = -1$, $z_3 = 1$, and $z_4 = m^{-1/2}$, where m is a parameter that represents the degree of freedom in the prevertices. The image of infinity turns out to be the point iK', and the image of 0 is 0. See Figure 2.11. The mapping function can

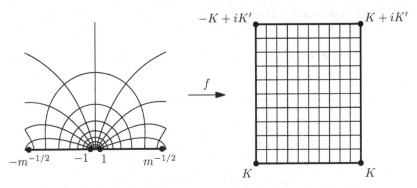

Figure 2.11. Map (2.11) to a rectangle ($n = 4$) from the upper half-plane.

Figure 2.12. Map of a generalized quadrilateral to a rectangle of unique aspect ratio. Two vertices are indistinguishable in the middle picture.

be written as an elliptic integral of the first kind:

$$w = f(z) = A + C \int^z \prod_{k=1}^{4} (\zeta - z_k)^{-1/2} \, d\zeta$$

$$= C \int_0^z \frac{d\zeta}{\sqrt{(\zeta^2 - m^{-1})(\zeta^2 - 1)}}$$

$$= C \int_0^{\sin^{-1} z} \frac{d\theta}{\sqrt{1 - m \sin^2 \theta}}. \tag{2.11}$$

Thus we have $\operatorname{sn}(C^{-1}w \mid m) = z$, where $\operatorname{sn}(u \mid m)$ is the **Jacobi elliptic sine** with parameter m.[1] If the rectangle is scaled so that $K = f(1)$ is the complete elliptic integral of the first kind with parameter m, our normalization ensures that $C = 1$, and the inverse map is simply $z = \operatorname{sn}(w \mid m)$. In this case K' is the complete elliptic integral with parameter $1 - m$. (Elliptic functions arise again in section 4.9. For more about their many properties, see [Hil59].)

The elliptic parameter m is associated in a one-to-one manner with the geometry of the rectangle—specifically, with its aspect ratio, since that is the only feature preserved by scaling, rotation, and translation. Thus the SC parameter problem is in principle solved by finding the value of m such that $K'(m)/2K(m)$ is the aspect ratio.

This rectangle transformation is closely associated with the notion of a **generalized quadrilateral**. A generalized quadrilateral Q is a Jordan region (i.e., bounded by the image of the unit circle under a continuous, one-to-one function) together with four distinguished points a, b, c, d lying in order on the boundary. A conformal transformation of the region to H^+ maps a, b, c, d to points in order on the real line. Those points can be mapped to ± 1 and $\pm m^{-1/2}$ by Möbius transformation for a unique value of m, and thence by an SC map to a rectangle of unique aspect ratio (see Figure 2.12). This aspect ratio is known

---

[1] The elliptic modulus k is sometimes used in place of m. Their relationship is $m = k^2$.

as the **conformal modulus** or conformal module of Q. Note that it depends on the four selected boundary points as well as on the geometry of the region. Two generalized quadrilaterals are conformally equivalent if and only if they have the same modulus. Practical implementation of SC mapping for the generalized quadrilateral problem is discussed in section 4.3.

2.6 Crowding

The rectangle map provides a canonical illustration of a fundamental phenomenon in conformal mapping known as **crowding**. Crowding is a form of ill-conditioning that causes trouble in virtually all numerical methods for conformal mapping. The first discussion of crowding in the literature appears to have been by Gaier [Gai72], and the term seems to have been invented by Zemach [MZ80]. Its relevance to SC mapping was highlighted in [Tre80].

The situation can be exemplified by the map from the disk to a rectangle (Figure 2.13). By conformality, the angles at which the curves meet at the origin are the same as in the disk. As the aspect ratio grows, the angles between some pairs of these curves become exponentially small. This effect can be analyzed via the expansions

$$K = \frac{\pi}{2} + O(m),$$

$$K' = \log \frac{4}{m^{1/2}} + O(m \log m).$$

$$a = 1$$

$$a = 4$$

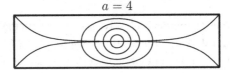

$$a = 7$$

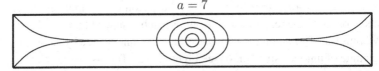

Figure 2.13. Illustration of the "crowding" phenomenon in maps of the disk to rectangles of different aspect ratio a. The angles subtended at the origin are $\pi/2$, 0.00475, and 0.0000427.

The aspect ratio a thus satisfies

$$a = \frac{K'}{2K} = \frac{1}{2\pi} \log \frac{16}{m} + O(m \log m),$$

$$m = O(e^{-2\pi a}). \tag{2.12}$$

The cross-ratio (see p. 31) of the four prevertices $-m^{-1/2}, -1, 1, m^{-1/2}$ is therefore $O(m^{1/2}) = O(e^{-\pi a})$. Since any map of the upper half-plane to the disk preserves this cross-ratio, some pair of prevertices in any disk map to the rectangle of aspect ratio a will be separated by a distance that is $O(e^{-\pi a/2})$.

In the vicinity of those crowded prevertices, the derivative of the map $f(z)$ must be $O(e^{\pi a/2})$. This indicates that tiny changes in a point in the disk—such as those induced by roundoff error on a computer—can be amplified enormously by the map. To put it in other terms, one generally cannot even distinguish adjacent prevertices in standard IEEE double precision arithmetic (roundoff approximately 2×10^{-16}) if a is larger than about 23.

Mapping from the half-plane might seem to alleviate crowding, because prevertices are separated by distances of size $O(1)$ or $O(m^{-1/2})$, neither of which is small. However, the *range* of magnitudes in the prevertex spacings remains, and this is what floating-point arithmetic limits.

Crowding is not limited to rectangle maps. Indeed, it occurs whenever the target region has areas that are relatively long and thin. We informally call such regions **elongated**. One interpretation of crowding is that whereas boundary data for an elliptic partial differential equation influence the solution throughout the region, this influence may be exponentially weak in areas separated from the point by an elongation. A common response to this difficulty in other elliptic settings is to use a domain decomposition; we shall consider this idea in section 3.4 (see also [Lau94, PS91, FPS99]). Another, more specialized circumvention of the problem is to give up the half-plane and disk as fundamental domains for elongated regions; we shall take this approach in section 4.3 (see also [How94]).

3

Numerical methods

Methods for the numerical computation of Schwarz–Christoffel maps have been refined for 40 years, and improvements continue. At this writing, it is possible to compute (to, say, eight accurate digits) maps to regions with one or two dozen vertices in a few seconds on a workstation. Regions with a hundred vertices can be treated in a few minutes.

In this chapter we shall examine the major algorithmic issues of SC mapping and refer to the literature for the full numerical analysis. For concreteness we describe maps from the disk, which is often most convenient computationally; the half-plane is little different. The chapter closes with brief descriptions of relevant software packages available for free in the public domain.

3.1 Side-length parameter problem

At the heart of any numerical method for Schwarz–Christoffel maps is the solution of the parameter problem. The most successful general-purpose methodology, based on equations derived from side lengths, was introduced in 1980 by Trefethen [Tre80], who built on work by Reppe [Rep79] and others. Variations on this technique remain the mainstay of numerical SC mapping. (A quite different method, based on cross-ratios and designed to overcome crowding, was proposed by Driscoll and Vavasis [DV98] and is explained in section 3.4.)

We recall the SC formula (2.4):

$$f(z) = A + C \int^z \prod_{k=1}^n \left(1 - \frac{\zeta}{z_k}\right)^{\alpha_k - 1} d\zeta. \tag{3.1}$$

The exponents in the integrand induce the correct angles in the image of the unit disk, regardless of where the prevertices lie on the unit circle. However, the locations of the prevertices determine the side lengths of the resulting image, as illustrated in Figure 1.3. In order to map to a given target, then, we must determine the locations of the prevertices by enforcing conditions involving the side lengths.

The three degrees of freedom in the map may be pinned down by specifying $z_{n-2} = -1$, $z_{n-1} = -i$, and $z_n = 1$. We are left with $n - 3$ real quantities—the arguments of the remaining prevertices—to determine. For a bounded polygon, this is accomplished by the $n - 3$ real conditions

$$\frac{\left| \int_{z_j}^{z_{j+1}} f'(\zeta)\, d\zeta \right|}{\left| \int_{z_1}^{z_2} f'(\zeta)\, d\zeta \right|} = \frac{|w_{j+1} - w_j|}{|w_2 - w_1|}, \qquad j = 2, 3, \dots, n - 2, \qquad (3.2)$$

where f' comes from (3.1). The following theorem explains why vertex w_n does not explicitly appear in these conditions.

Theorem 3.1. *Assume that $\alpha_n \neq 1$ and $\alpha_n \neq 2$. A bounded polygon is uniquely determined, up to scaling, rotation, and translation, by its angles and the $n - 3$ side-length ratios appearing on the right-hand sides of (3.2).*

Proof. Because we have scaling, rotation, and translation available, we can assume that w_1 and w_2 are correct. Because $|w_3 - w_2|$ and the angle of the polygon at w_2 are known, the location of w_3 is determined. Proceeding inductively, we can position all the vertices through w_{n-1} in the same manner. To locate w_n, we note that the lines containing the sides that adjoin w_n are not parallel, because of the conditions on α_n. Since the angles at w_1 and w_{n-1} are known, we can locate w_n at the unique intersection of two nonparallel lines. See Figure 3.1. □

The conditions on α_n are not restrictive in practice, as we are free to number the vertices so that a convenient vertex is the nth. The scaling, rotation, and

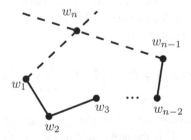

Figure 3.1. Location of the last vertex by intersection.

translation are possible because of the constants A and C appearing in (3.1). Conveniently, however, neither constant appears in (3.2).

If $w_J = \infty$ for $J < n$, two of the conditions in (3.2) are meaningless. We can replace them by the complex condition

$$\frac{\displaystyle\int_{z_{J-1}}^{z_{J+1}} f'(\zeta)\,d\zeta}{\left|\displaystyle\int_{z_1}^{z_2} f'(\zeta)\,d\zeta\right|} = \frac{w_{J+1} - w_{J-1}}{|w_2 - w_1|}. \tag{3.3}$$

This condition ensures that the two components of the boundary that are incident on w_J are positioned correctly with respect to each other. As a consequence, we must require that no two infinite vertices be adjacent, because otherwise (3.3) is still useless. If this requirement is not met initially, we can introduce one degenerate vertex with interior angle π on the straight line between infinite neighbors.

We now have $n - 3$ real side-length conditions on the unknown prevertices, staring with (3.2) and substituting as needed from (3.3). It is crucial when solving for the prevertices that they be constrained to lie in order on the unit circle. Designating the argument of z_k by θ_k, we let

$$0 < \theta_1 < \theta_2 < \cdots < \theta_n = 2\pi. \tag{3.4}$$

Our earlier choices for the three degrees of freedom imply also that $\theta_{n-2} = \pi$ and $\theta_{n-1} = \frac{3\pi}{2}$. Because constrained systems of equations may be difficult to solve (especially in light of crowding, as discussed in section 2.6), it is desirable to formulate an equivalent unconstrained system. One approach that has proved very successful in practice is the transformation

$$\phi_k = \log\left(\frac{\theta_k - \theta_{k-1}}{\theta_{k+1} - \theta_k}\right), \qquad k = 1, \ldots, n - 3.$$

Here we adopt the convention $\theta_0 = 0$. The ϕ_k variables take arbitrary real values, and the relationship is easily inverted:

$$p_0 = 1,$$

$$p_k = \prod_{j=1}^{k} e^{-\phi_j}, \qquad k = 1, \ldots, n - 3,$$

$$\theta_m = \pi \frac{\sum_{k=0}^{m-1} p_k}{\sum_{k=0}^{n-3} p_k}, \qquad m = 1, \ldots, n - 3.$$

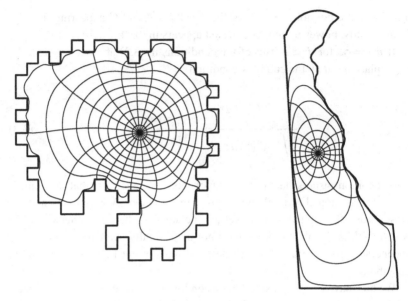

Figure 3.2. Maps to two regions with 100 vertices. The circular arc at the top of the state of Delaware (right) is approximated by 36 straight-line segments. In practice we recommend instead methods designed for smooth arcs.

The parameter problem has now been expressed as a system of equations in the unconstrained variables $\phi_1, \ldots, \phi_{n-3}$. This system must be solved by an iterative numerical method. Note that the Jacobian of the system is hard to express analytically. In most cases a quasi-Newton iteration [DS96] or other "black-box" solver will give the best results. However, the complexity of such methods is $O(n^3)$ as $n \to \infty$. In practice, a hundred vertices or so present no trouble, as illustrated in Figure 3.2. For very large n, though, simple linear iterations such as those in [Cos87, CS92, Dav79, FA74] might be useful, especially during the early iterations far from the solution. (However, if the large n arises from the approximation of smooth arcs, we recommend instead the use of methods specifically designed for such regions. See sections 3.6, 4.10, and 4.11.)

A theoretical difficulty can arise in the solution of these equations [How90]. Consider, for example, the polygonal regions of Figure 3.3. Deforming the region on the left continuously into that on the right would require the two slits to move past each other. However, if the slits are shortened or nearby sides are otherwise adjusted accordingly, any reasonable residual measure of side-length accuracy will increase. Assuming the use of a purely descending nonlinear equations method, iterations will therefore push the slits together. As the gap

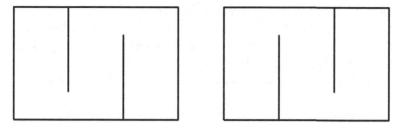

Figure 3.3. A descent method for the side-length parameter problem can never achieve the polygon on the right when starting from the polygon on the left.

between them shrinks toward zero, some prevertices in the disk become arbitrarily crowded, causing at least one ϕ_k variable to approach infinity.[1] Therefore, a descent method starting from an initial prevertex arrangement that maps to the region on the left will not converge to the region on the right—some sort of global "leap" is needed to allow the slits to cross. In practice, a breakdown like this is extremely rare, but regions with many bumps or slits can lead to very slow convergence unless supplied with fairly good initial guesses.

Once the prevertices have been found, the multiplicative constant C in (3.1) can be found by integrating between two prevertices. The additive constant A is simply the image of the base point of the integral; rather than fixing this point, it is most convenient to compute $f(z)$ by starting from the prevertex (among those whose images are finite) that is closest to z.

3.2 Quadrature

Schwarz–Christoffel mapping requires the rapid and accurate computational approximation of integrals of the form

$$\int_a^b \prod_{k=1}^n \left(1 - \frac{\zeta}{z_k}\right)^{\alpha_k-1} d\zeta.$$

Reproducing the relatively simple Figure 1.4, for example, took 8,249 such integrations (see the appendix), and the left side of Figure 3.2 required 94,204 of them. Because the integrand is analytic throughout the canonical mapping region, the integral is path-independent. We typically choose to use a straight-line segment, but in certain circumstances other choices are more appropriate.[2]

---

[1] From an algebraic standpoint, this is not the same as a nonglobal local minimum in the residual, which is impossible [How90].

[2] For example, in the exterior map of section 4.4 it is important not to choose a path that goes near the pole at the origin.

Because of the need to solve the parameter problem (see section 3.1), it often happens that one or both of the points a, b is a prevertex. Recalling that for each k, $-2 \leq \alpha_k \leq 2$, we see that the integrand in such cases lacks regularity at one or both ends unless $\alpha_k = 0, 1, 2$. These singularities presents a challenge for general-purpose integration methods, such as those based on Newton–Cotes rules.

The specific form of the singularity, however, allows us to apply **Gauss–Jacobi** quadrature [GW69]. This is a highly accurate method for integrals of the form

$$\int_{-1}^{1} r(t)(1-t)^{\beta}(1+t)^{\gamma} \, dt,$$

for smooth functions $r(t)$. By linearly rescaling the original interval $[a, b]$ to $[-1, 1]$ and dividing out the singular behavior, we obtain a smooth integrand of the proper form. (Newton–Cotes rules can also be modified to incorporate the singularities explicitly, in which case they perform acceptably [FZ88].)

However, a subtle problem remains. The Gauss–Jacobi formula accounts for singularities at the endpoints. When these endpoints are prevertices, the crowding phenomenon (section 2.6) implies that other prevertices may be extremely close to the interval of integration. Such singularities will severely degrade the rate of convergence of the quadrature.[3] To combat this problem, Trefethen [Tre80] proposed using a *compound* Gauss–Jacobi method, in which the integration interval is subdivided in accordance with the "one-half rule":

> **The one-half rule:** No singularity may lie closer to an integration subinterval than one-half the length of that subinterval.

Trefethen found that, as a rule of thumb, the number of accurate digits obtained by computing SC integrals with this compound method was roughly equal to the number of Gauss–Jacobi nodes used per subinterval.

This one-half rule is analogous to techniques used in hp finite element methods for the numerical solution of partial differential equations (for example, see [Sch98, pp. 89–100], and references therein), where the geometric subdivision idea goes by the name of geometric mesh refinement; similar ideas also appear in the literature of the numerical solution of integral equations. In particular, Scherer and Babushka and others have argued that, for a wide class of problems involving geometric mesh refinement, a refinement ratio of

---

[3] We are referring here to distances like 10^{-8}, not 10^{-1}, and the degradation of accuracy is not merely annoying but dominant.

$\sigma = (\sqrt{2} - 1)/2 \approx 0.17$ is optimal. This ratio would correspond to changing the number $1/2$ in the one-half rule to $1/A$ with $A = \sigma^{-1} - 1 \approx 4.8$. Experiments by Lehel Banjai (unpublished) indicate that this change improves performance of the SC Toolbox in some problems by a modest amount, on the order of 10%.

Other techniques for SC integration have been tried with some success [Cos87, Dav79, FZ88, KK64]. An early survey and comparison was done by Meyer [Mey79]. Howell [How90] studied various adaptive and singularity removal quadrature techniques and recommended the compound Gauss–Jacobi method.

In solving the parameter problem, what is really desired is a side length, that is,

$$\left| \int_{z_j}^{z_{j+1}} \prod_{k=1}^{n} \left(1 - \frac{\zeta}{z_k} \right)^{\alpha_k - 1} d\zeta \right|.$$

Krikeles and Rubin [KR88] pointed out that the absolute value can be taken inside the integral if the path is along the unit circle. As a result, the integrand can be evaluated using only real, not complex, logarithms—a considerable savings in practical computation. In the half-plane, such integrations are along the real axis and trivially also use only real logarithms. This special structure is lost in both cases, however, when a prevertex of infinity causes a substitution in the side-length conditions of the form (3.3), as described in section (3.1).

3.3 Inverting the map

Once the Schwarz–Christoffel parameters are known, evaluation of the forward map $w = f(z)$ is simply a matter of computing the SC integral (3.1). Inversion of the map to produce $z(w)$ is more difficult, because no formula exists in general. Trefethen [Tre80] proposed two strategies for inversion:

1. Newton iteration on the forward map $f(z) - w = 0$, and
2. Numerical solution of the initial-value problem (IVP)

$$\frac{dz}{dw} = \frac{1}{f'(z)} \quad \text{and} \quad z(w_0) = z_0.$$

The Newton iteration is attractive because $f'(z)$ is known (in fact, it is much cheaper to compute than $f(z)$ is) and convergence will be locally quadratic. However, in practice one may need a rather good starting guess to avoid divergence. Solving the initial-value problem, on the other hand, is more reliable but considerably slower. Trefethen recommended using a standard IVP solver with a fairly relaxed error tolerance to obtain an initial estimate for the Newton iteration. This is still the best method known.

There is a practical difficulty in identifying w_0 and z_0 to begin the IVP. The only points at which the inverse map is known after solving the parameter problem are the vertices, but the differential equation is generally singular at those points. Moreover, it is easiest to solve the IVP along a straight line from w_0 to the desired point w, but then one must choose w_0 so that this path lies entirely inside the polygon. An inelegant approach that is nonetheless successful in software is to compute the forward map at points on the unit circle between prevertices and exhaustively check the path condition. Typically the work involved with this step is insignificant compared to the IVP/Newton steps.

3.4 Cross-ratio parameter problem

As was noted in section 2.6, any elongated areas in the target region force prevertices to be placed extremely close together, with separations that shrink exponentially with the local aspect ratio. This crowding causes the derivative $f'(z)$ to be large near those prevertices, which in turn leads to great sensitivity of the values of the map to roundoff error. In the most extreme circumstance, the exact prevertices cannot be distinguished in the floating-point representation of the computer.

One algorithmic response to crowding has been to change the fundamental domain. The prime example of this technique is the use of a strip or rectangle to map to regions that are elongated in just one major sense—say, an L-shaped or Z-shaped region (see sections 4.2 and 4.3). This strategy reflects the view that the disk or half-plane is an ill-conditioned choice of domain for such a target region, and it is both reasonable and successful when applicable. However, neither the strip nor the rectangle is suitable for a T-shaped domain. It is possible to construct the SC map from a slit strip for such domains [How94], but the formula becomes more complicated. Furthermore, even though formulas exist or can be derived for H-shaped or arbitrarily elongated domains, they rapidly become unwieldy, and automatic selection of the proper formula for a particular target domain is difficult.

As a first step toward a different response to crowding, consider the situation in Figure 3.4. Recall that the prevertices have three degrees of freedom among them; that is, there is a three-parameter family of prevertex arrangements that all map to the same polygon. (The constants A and C in (2.4) depend on the free parameters.) We call each such prevertex arrangement an **embedding**. When the target region is highly elongated, each embedding will lead to crowding

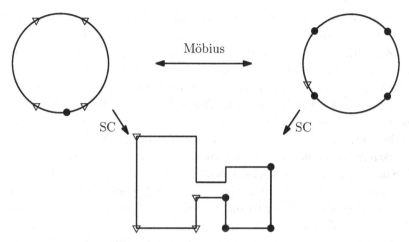

Figure 3.4. Both embeddings of the disk (top) map to the same polygon (bottom), but different groups of prevertices (dots and triangles) are clustered on the circle. Hence each version of the map is crowded in a different part of the polygon.

somewhere in the map. But this crowding may be highly localized, and the location may be different for different embeddings, as is the case in Figure 3.4.

The trouble in the side-length algorithm for an elongated polygon originates from fixing the three free parameters. A particular embedding is implicitly chosen at the start, and it must suffer from crowding somewhere. To circumvent crowding, we need a robust way to find the *family* of equivalent embeddings. More specifically, we need

1. A compact representation of the family, and
2. Access to locally well-conditioned embeddings from the family.

We begin with the first item. Any representation of an embedding family should have $n - 3$ real components. Moreover, because each embedding is related to any member of its family by a conformal map of the disk to itself (Möbius transformation), the family representation must be invariant under such maps. A good choice for the representation is to define $n - 3$ ordered 4-tuples of prevertices and take the **cross-ratio** of each, defined as [Neh52]

$$\rho(a, b, c, d) = \frac{(d - a)(b - c)}{(c - d)(a - b)}. \tag{3.5}$$

It is easily seen that for four points in counterclockwise order on the unit circle (or, in fact, any circle), the cross-ratio is real and negative. Cross-ratios are also invariant under Möbius transformations.

How should we choose the $(n - 3)$ prevertex 4-tuples? One obvious choice is to take the consecutive 4-tuples on the indices $(1, 2, \ldots, n)$. However, a geometrically motivated choice has important advantages. The result is the **CRDT** algorithm (cross-ratios of the Delaunay triangulation) [DV98].

We restrict attention to bounded polygons. Let $\mathcal{T}$ be the constrained Delaunay triangulation [BE92] of the polygon vertices. (The Delaunay triangulation is optimal in the sense of maximizing the minimal interior angle.) The triangles of $\mathcal{T}$ have sides that include segments drawn through the interior of P; we call such sides **diagonals** of $\mathcal{T}$. Elementary facts about triangulations on n vertices include that there are $n - 2$ triangles in $\mathcal{T}$ and $n - 3$ diagonals. Each diagonal is a side for two triangles in $\mathcal{T}$, and the union of those triangles is a quadrilateral. We denote the set of $n - 3$ quadrilaterals by $\mathcal{Q}$. The vertices of each $Q \in \mathcal{Q}$, taken in counterclockwise order, define an ordered 4-tuple of prevertices. See Figure 3.5. Observe also that the quadrilaterals can be described by a connected "overlap graph" in which adjacency is defined by the sharing of a triangle.

We now propose a representation of an embedding family. For each quadrilateral $Q \in \mathcal{Q}$, let $\sigma(Q)$ be the cross-ratio of that quadrilateral's prevertices. There are $n - 3$ such parameters, and they are invariant under the Möbius

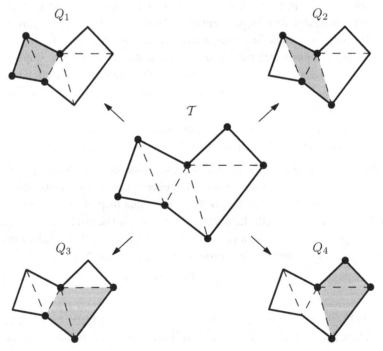

Figure 3.5. Triangulation $\mathcal{T}$ and quadrilaterals $Q_1, \ldots, Q_4$ in the CRDT scheme for $n = 7$. Each quadrilateral defines a 4-tuple of prevertices, the cross-ratio of which is used as an unknown in the parameter problem.

(equivalence) transformations of the family. We need to show that, conversely, each collection of σ values does in fact correspond to a family of embeddings. We do this constructively and in a manner that leads into item #2 in the preceding list—the ability to extract particular, locally well-conditioned embeddings.

It is easily seen that if the four prevertices of Q are constrained to form a rectangle on the unit circle, then they are uniquely determined (up to a trivial rotation) by $\sigma(Q)$. Now, Q shares a triangle with some other quadrilateral, Q'. (It may help to refer again to Figure 3.5.) Hence we know three of the prevertex positions for Q', as well as $\sigma(Q')$. This information uniquely determines the fourth prevertex of Q'. By visiting quadrilaterals according to the overlap graph (say, in depth-first order), we can construct a complete embedding of prevertices in this fashion. Since the process began with Q and clearly depends on the choice of Q, we call this embedding $E(Q)$.

We have now satisfied the first item of our list. The second is also under control, thanks to the following property:

Well-separation property: If $\max_{Q \in \mathcal{Q}} |\log(-\sigma(Q))|$ is not large, each of the prevertices of Q in $E(Q)$ is well-separated from all the others.

The reasoning here is simple, if imprecise. The prevertices of Q in $E(Q)$ form a rectangle. If two sides of the rectangle were separated by a small ϵ, then $\sigma(Q)$ would be $O(\epsilon^{\pm 2})$. That would contradict the assumption, so the initial four prevertices are mutually well-separated. Suppose that Q' overlaps with Q and that the new prevertex it introduces is within ϵ of a prevertex of Q. Only one of the four distances needed to define the cross-ratio $\sigma(Q')$ can be small, so then $\sigma(Q') = O(\epsilon^{\pm 1})$, another contradiction. The argument proceeds inductively, with each new prevertex forbidden to be close to any prevetex of Q. In practice, by well-separated we mean agreement to at most three or so digits, thereby presenting no problem for double precision arithmetic.

An important aspect of the well-separation property is that in $E(Q)$ it applies *only* to the prevertices of Q. Other prevertices could be close to each other—indeed, this is generally the case. However, for purposes of evaluating the SC integral, this fact is immaterial as long as the interval of integration is not near such clusters. In other words, $E(Q)$ is a suitable embedding for computing values of the SC map that end up in or near Q. The situation is summarized in Figure 3.6 for a polygon with $n = 7$ vertices and hence four quadrilaterals, labeled $Q_1, \ldots, Q_4$. Each embedding $E(Q_j)$ has the prevertices of Q_j well-separated and arranged in a rectangle. These four prevertices define all of $E(Q_j)$ and hence an SC integral, $I_j(z)$. The image of the disk under I_j is a scaled,

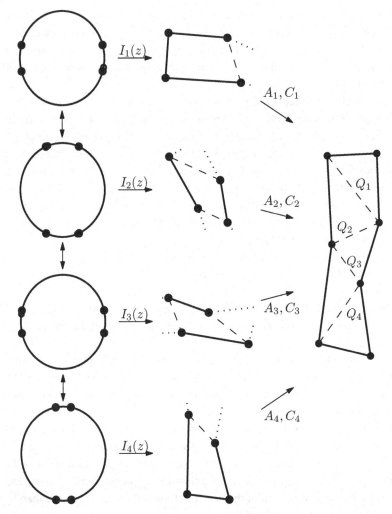

Figure 3.6. Summary of the key CRDT idea. Each prevertex embedding on the left defines a particular SC integral, denoted $I_1, \ldots, I_4$. The distinguished rectangle of prevertices (solid dots) map to a 4-tuple that can be scaled and rotated using the affine constants A_1, C_1, etc., to align with quadrilaterals $Q_1, \ldots, Q_4$ in the given polygon on the right. Remaining prevertices (open circles) are crowded, but that is immaterial.

rotated, and translated copy of the target region. Thus a pair of affine constants A_j, C_j complete the map from $E(Q_j)$ to the target.

At this point we need to ask how to ensure the condition that leads to the well-separation property. We know that the statement "$\log(-\sigma(Q))$ large" suggests that Q must be rather elongated. To prevent large or small cross-ratios, then,

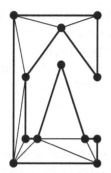 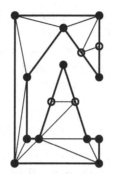 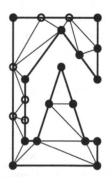

Figure 3.7. Two-stage preprocessing step for the CRDT algorithm. Vertices are first added to the original polygon (left) in order to cut off sharp corners (center). Then, additional vertices are added to break up long, narrow channels (right).

we introduce a preprocessing step for the polygon Γ whose purpose is to bound the effective aspect ratio of every quadrilateral in $\mathcal{Q}$. This goal is achieved by introducing degenerate vertices with interior angle π along the sides of Γ. The details of one such technique are presented in [DV98] and are illustrated in Figure 3.7. In practice, the number of vertices in Γ can grow substantially, adding to the size of the representation of embeddings; however, at least the degenerate vertices do not appear in the SC integrand.

The prevertex cross-ratios that produce the correct embedding family must be found by solving a parameter problem. The $n-3$ necessary conditions are determined geometrically. One could again use the side-length conditions (3.2); each side of Γ belongs to a quadrilateral, so there is at least one embedding in which its length can be computed reliably. However, the CRDT algorithm uses an alternative set of conditions that seems to lead to more easily solved systems. Given a list of prevertex cross-ratios, each quadrilateral $Q \in \mathcal{Q}$ is considered in turn. The prevertices of Q in $E(Q)$ are mapped under the SC integral to form an approximation $\tilde{Q}$ to Q (refer to the middle column of Figure 3.6). One condition is derived by requiring the moduli of the cross-ratios of Q and $\tilde{Q}$ to agree. (This number is independent of the affine scaling constants, which therefore do not need to be computed for solving the parameter problem.) Since there are $n-3$ quadrilaterals in $\mathcal{Q}$, this defines $n-3$ real conditions. Snoeyink [Sno99] proved that these conditions (plus the known angles) uniquely determine a polygon.[4]

The cross-ratios $\{\sigma(Q): Q \in \mathcal{Q}\}$ that determine the embedding family are real and negative, hence constrained. We negate them and take a logarithm to

---

[4] Technically, it is not clear whether Snoeyink's proof applies to every iteration in a solution of the parameter problem, since some embeddings yield self-intersecting "polygons" that do not meet the usual definition. In practice there seems to be no difficulty.

make them unconstrained. For scaling purposes, it makes sense to take logarithms of the geometric conditions as well. The resulting system of nonlinear equations can be solved by a quasi-Newton iterative method. Experimentally we find that the system is rather well-behaved, and more economical Jacobian strategies are useful; see [DV98] for details.

3.5 Mapping using cross-ratios

Evaluating the SC map using the CRDT formulation is less straightforward than in the case where prevertices are found directly. Each embedding has an associated pair of affine constants that complete the SC map to the target region (see Figure 3.6). These can be computed directly once the proper embedding family has been found. We also need to be able to relate any two of the computational embeddings. To this end, note that equivalent embeddings are related by self-maps of the disk (shown as the vertical arrows on the left-hand side of Figure 3.6), which are in turn determined by the mapping of three points. Numerical instability would arise if we chose three points in the source embedding that were crowded in the destination embedding. Fortunately we can avoid this situation. Two overlapping quadrilaterals Q and Q' in $\mathcal{Q}$ share three prevertices, and these prevertices are well separated in both $E(Q)$ and $E(Q')$. So we must always transform between neighboring embeddings in the overlap graph; a direct transition from the top left of Figure 3.6 to the bottom left would numerically fail.

To compute values of a map, one must choose a "reference" embedding in which source points are given. We require that the reference embedding be specified by the locations of three prevertices corresponding to a triangle. Then it is readily transformed to one of our computational embeddings, and thence to any other.

Suppose that a point z in the reference embedding is to be mapped to the target region P. Since $f(z)$ is inside at least one quadrilateral, we know that at least one CRDT embedding is suitable. However, it is impossible to know a priori what the best embedding may be. (Inverse mapping is, in this regard, simpler.) Hence we map z from the reference to *all* of the other embeddings, since this is computationally inexpensive. We heuristically expect that the best embedding is the one in which the transformed z is closest to the origin, and we use it for computing the map.

Figure 3.8 illustrates the power of the CRDT method in mapping to a multiply elongated "maze." Cross-ratios describe the disk map, which in turn is used as an intermediary to a conformally equivalent rectangle. Two opposite sides of the rectangle map to the "entrance" and "exit" of the maze, and four paths

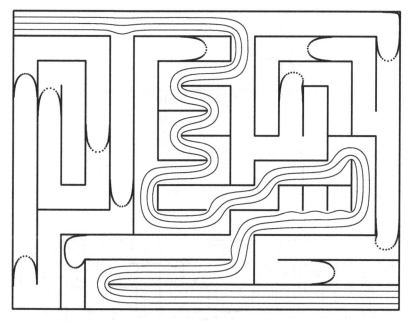

Figure 3.8. Map to a "maze" using the CRDT cross-ratio formulation. Each curve is the image of a line in a conformally equivalent rectangle; the dotted curves represent lines which are exponentially close (10^{-10}, 10^{-20}, ..., 10^{-50}) to a rectangle side. All computations were done using double precision arithmetic.

through the maze (solid curves) are the images of straight lines in the rectangle. The dots shown also lie on images of parallel lines—in this case, lines that are exponentially close (10^{-10}, 10^{-20}, ..., 10^{-50}) to a side of the rectangle. For most of their length, they are indistinguishable from the maze walls, until one penetrates well into a "dead end." The results are accurate despite the fact that the computations are performed in double precision arithmetic (machine epsilon $\approx 10^{-16}$). We are not aware of any other technique that can accurately find these dots using only double precision arithmetic.[5]

Another sort of map to the same region is illustrated in Figure 3.9. Instead of using the disk as an intermediary to a rectangle, the canonical region is multiply elongated to better match the "arms" of the maze. All the angles of this domain are multiples of $\pi/2$, which makes it simple to construct an orthogonal grid. The side lengths in this domain are not known a priori but must be found as part

---

[5] We note here that the original maze has 120 vertices, which became 180 vertices after the CRDT preprocessing subdivision. The cross-ratios were found in about 13 minutes using the SC Toolbox on a 2000-vintage workstation. Plotting the dots in Figure 3.8 took about 10 lines of MATLAB code plus some human intervention.

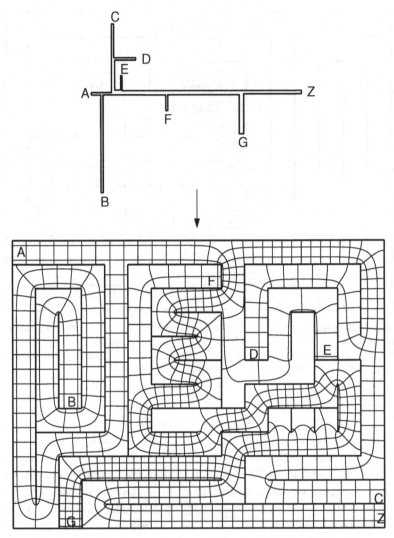

Figure 3.9. Map to the maze from a multiply elongated region. The maze (bottom) is first SC mapped to the disk. The resulting prevertices are assigned angles that are multiples of $\pi/2$. When side lengths are computed from the SC integral, the region at top results. A rectangular grid on this region (not shown) is then mapped to the maze. (Letter labels are used to aid comparison of the two regions.) All this is done accurately by CRDT in double precision arithmetic.

of the computation by evaluation of the SC integral; this is a generalization of the conformal modulus problem.

3.6 Software

The most comprehensive and user-friendly numerical SC software is the Schwarz–Christoffel Toolbox for MATLAB. For more on the toolbox, how to obtain it, and examples of its use, see the appendix. Except where explicitly noted, all the examples and figures in this book were computed using this toolbox. (Some require a fair amount of programming in addition to the distributed toolbox code, however.)

The SC Toolbox is a descendent of SCPACK, a FORTRAN 77 package written by Trefethen [Tre80, Tre89]. SCPACK was responsible for first making numerical SC mapping widely available to nonexperts and has been in use for 20 years. Only maps from the disk, using the side-length formulation of the parameter problem, are directly supported; routines are included for computing the forward and inverse maps. SCPACK is available through the Netlib repository at http://www.netlib.org/conformal.

Also at Netlib in FORTRAN are:

DSCPACK A package written by Hu [Hu95] for mapping from annuli to doubly connected regions bounded by polygons (see section 4.9). As in SCPACK, solving the parameter problem and computing maps in either direction are supported.

CAP A package by Bjørstad and Grosse [BG87] for mapping to circular-arc polygons (see section 4.10). This package is described by the authors as "experimental" and is not as robust as others listed here.

GEARLIKE A package written by Pearce [Pea91] for maps to "gearlike" domains (see section 4.8). This is modeled after SCPACK.

Finally, we mention three other public-domain packages for non-SC-based conformal mapping:

CONFPACK A FORTRAN package by Hough [HP83, Hou90] which implements Symm's equation for simply connected regions (interior or exterior) with piecewise smooth boundaries. CONFPACK explicitly accounts for corner singularities [Hou89] and is very accurate and efficient. It is available at Netlib.

zipper A C package written by Marshall that implements an interpolation-based method of Kühnau [Küh83] for interior and exterior regions. Although

corners are not explicitly accounted for, the software is remarkably fast.
It is available from its author at `http://www.math.washington.edu/~`
`marshall/zipper.html`.

CirclePack A C package written by Stephenson that computes packings of
circles with specified tangency. These can be used as approximations to conformal maps. See [RS87, Ste99] and `http://www.math.utk.edu/~kens`
for more details.

4

Variations

A key aspect of the power of the Schwarz–Christoffel transformation (indeed, a large part of the motivation for this book) is its remarkable flexibility in adapting to a wide variety of situations, not all of which superficially seem to involve conformal maps or even polygons. The essence of SC mapping is to treat the corners exactly; if the rest of the problem is simple, nothing else is needed. What emerges from applications of this principle is that Schwarz–Christoffel mapping is not just a mapping technique but a distinctive way of thinking about problems of potential theory in the plane.

Let us reconsider the fundamental SC philosophy for constructing a map $f(z)$. For the half-plane, we required f' to have piecewise constant argument along the boundary because then the image under f has straight lines with corners. For other canonical domains, we need to modify this requirement slightly. For example, as we follow the boundary of the unit disk, a constant argument for f' does not lead to a straight-line image. However, if $g(z)$ is a function that "straightens out" the original domain boundary, then f'/g' will have the appropriate piecewise-constant argument. This fact is especially convenient because we can also use powers of $g(z) - g(z_k)$ to create wedges that have the right jumps. To summarize, the map from D to polygonally bounded P satisfies the following rule:

Let f map D to the polygonally bounded region P (with interior angles $\pi\alpha_k$), and let g map the boundary of D to a straight line. Then

$$\frac{f'(z)}{g'(z)} = C \prod_{k=1}^{n} [g(z) - g(z_k)]^{\alpha_k - 1}. \tag{4.1}$$

41

In fact, (4.1) is the chain rule for differentiation of the function $f(z) = h(g(z))$, where h is a standard half-plane map. But our geometric interpretation of this formula can be useful in the derivation of other variations and applications. We shall see that maps in which more than one boundary component is present (sections 4.2 and 4.9) require refinement of the idea of "straightening" and some modification of (4.1).

We must recognize too that this form leads only to proper *local* behavior. A polygon has a special global property—namely, it winds about its interior once—and other aspects of the map may also require a global perspective. In such cases the elementary SC factors need to be given an additional adjustment (see section 4.4, for example).

4.1 Mapping from the disk

The unit disk may be preferred to the half-plane in some computations because certain boundary conditions are more naturally applied there. The disk also has the computational advantage of being a bounded domain and the aesthetic advantage of having no naturally distinguished boundary point. We have already encountered the disk formula in (2.4), but we derive it here for completeness. Throughout, we assume that the point -1 is not among the prevertices; we can easily arrange this situation by a rotation.

For our function $g(z)$ in the basic formula (4.1), we can choose the Möbius transformation $1/(1 + z)$, which maps the disk to a half-plane bounded to the left by the line $\text{Re}\, z = \frac{1}{2}$. (Our choice of $g(z)$ is not unique, and other choices would lead to equally valid formulas.) Thus,

$$
\begin{aligned}
f'(z) &= C g'(z) \prod_{k=1}^{n} (g(z) - g(z_k))^{\alpha_k - 1} \\
&= C(1 + z)^{-2} \prod_{k=1}^{n} \left(\frac{z - z_k}{1 + z} \right)^{\alpha_k - 1} \qquad (4.2) \\
&= C \prod_{k=1}^{n} (z - z_k)^{\alpha_k - 1}
\end{aligned}
$$

for some constant C.[1] The last step reflects the fact that the exponents sum to -2 by (2.1). As explained after (2.4), this formula is equivalent to the more numerically convenient

---

[1] Recall our convention that this constant may change from line to line to absorb extraneous factors.

Schwarz–Christoffel formula from the disk

$$f(z) = A + C \int^z \prod_{k=1}^{n} \left(1 - \frac{\zeta}{z_k}\right)^{\alpha_k - 1} d\zeta$$

Figures 4.1 and 4.2 present some examples of SC disk maps.

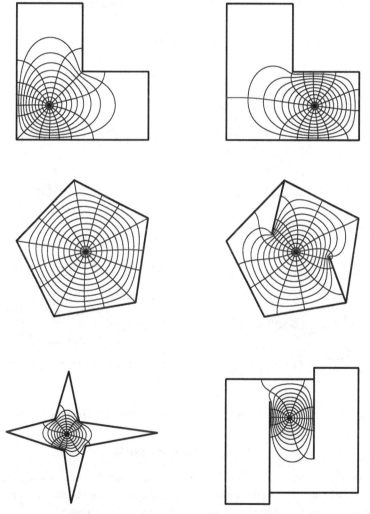

Figure 4.1. Examples of disk maps to bounded regions. In each case a regular polar grid in the disk is mapped to the target region. At bottom right, the prevertices on the unit circle are moderately crowded—their minimum separation is about 3.6×10^{-5}.

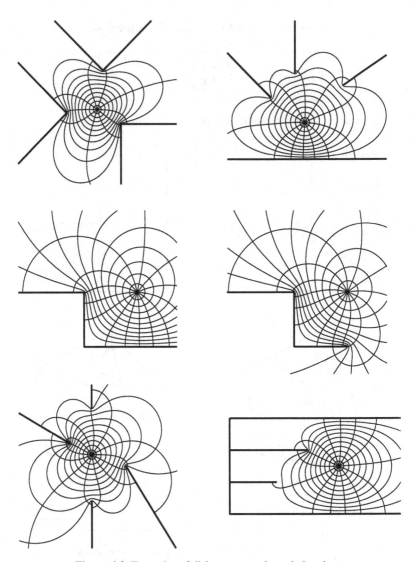

Figure 4.2. Examples of disk maps to unbounded regions.

4.2 Mapping from a strip

The infinite strip $S = \{0 < \operatorname{Im} z < 1\}$ is especially useful as a canonical domain, for two reasons. First, many applications involve an infinite channel, or Dirichlet boundary conditions with just two constant values (see section 5.2). Second, because S is essentially the logarithm of the upper half-plane, crowded prevertices on the real line may become well separated on the boundary of S and therefore

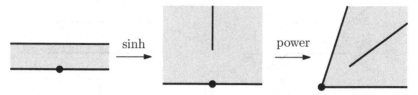

Figure 4.3. Elementary factor in the SC strip map. The argument of the image jumps at z_k and at the ends of the strip.

easier to treat numerically. As far as we know, this map was first described by Sridhar and Davis in [SD85] and implemented numerically in [Flo85]. It was implemented and advocated as a method for combating crowding in [HT90].

The basic formula (4.1) is not quite directly applicable. Our "straightening function" g is required to map the lower boundary Im $z = 0$ to a line through zero (giving a piecewise constant argument) and the upper boundary Im $z = 1$ to a radial slit (giving constant argument). But then translation by a lower prevertex, as suggested by (4.1), will cause variation of the argument on the formerly radial slit. Instead we translate *initially*, since that leaves the strip unchanged, and then apply g. This idea leads to the function

$$\sinh\left[\frac{\pi}{2}(z - z_k)\right], \tag{4.3}$$

which maps S to a slit half-plane. Taking a power transforms this region further into a slit wedge (see Figure 4.3). The slit wedge gives the proper jump in arg f' at z_k without affecting the argument elsewhere on the boundary of S. Due to the structure of the sinh function, the same factor (4.3) also works when z_k is on the upper boundary.

Each of the elementary factors also causes jumps in argument at the ends of the strip, which always map to infinity. The jump at each end is half of the wedge angle shown in Figure 4.3. If the sum of exponents at the prevertices is -2, then each end has an overall $\int z^{-1}\, dz$, or logarithmic, behavior of parallel sides extending to infinity. In other cases the divergences are nonparallel, as illustrated in Figure 4.4. However, under the map so far described, the two ends of the channel always have identical "divergence angles." To effect different divergences at the ends, we need to turn the image of the top side of S while leaving the other side fixed, as illustrated in the bottom row of Figure 4.4. A factor that will have this effect on f' is

$$\exp\left[-\tfrac{1}{2}\pi(\alpha_+ - \alpha_-)z\right].$$

Here we use $\pi\alpha_\pm$ to denote the desired divergence angles at the ends of the strip.

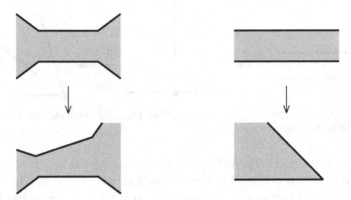

Figure 4.4. Divergence adjustment of the strip map. Before adjustment, the divergence angles at the two ends of the strip are identical (top row). By turning just one side of the image, though, we can create asymmetric divergence or even finite intersections (bottom row).

Putting the map together, we have

Schwarz–Christoffel formula from a strip

$$f(z) = A + C \int^z \exp\left[\frac{\pi}{2}(\alpha_- - \alpha_+)\zeta\right] \prod_{k=1}^{n}\left[\sinh\frac{\pi}{2}(\zeta - z_k)\right]^{\alpha_k - 1} d\zeta \qquad (4.4)$$

Some examples of strip maps are shown in Figures 4.5 and 4.6.

Because the two ends of the strip are preassigned, only one degree of freedom remains in the map: sideways translation of the strip, which can be fixed by placing a designated prevertex at zero. Thus $n - 1$ real conditions are needed to determine the remaining prevertices, where n is the number of vertices exclusive of the channel ends. Suppose that w_1 is the first vertex after the image of the left end of the strip and w_b is the first vertex before the image of the right end. (In other words, z_1 and z_b are the first and last prevertices on the bottom of S.) Using side lengths between adjacent pairs from w_1 to w_b and from w_{b+1} to w_n yields $n - 3$ conditions. (If some of these vertices are infinite, two side-length conditions must be replaced by one complex condition as described by (3.3).) The remaining two real conditions arise from the complex difference $w_1 - w_n$, which orients the images of the two sides of S with respect to each other.

As always, it is important for numerical computation to deal with branch cuts correctly. Using the principal branch in the form in (4.4) is fine for prevertices on the bottom side of S, but the argument of the power should first be negated for prevertices on the top side.

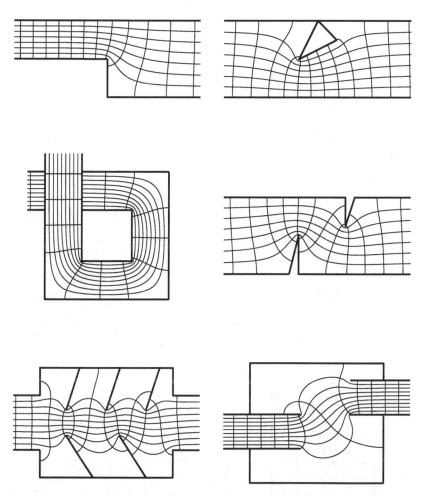

Figure 4.5. Examples of strip maps. In each case a regular cartesian grid in the infinite strip is mapped to the target region.

4.3 Mapping from a rectangle

Another important fundamental domain for SC maps is the rectangle. As with the strip, the motivations are both applied and numerical. Idealized electrical resistance, for example, is most easily computed (when possible) by transplantation to a rectangle, where the two pairs of opposite sides are assigned Dirichlet and homogeneous Neumann boundary conditions (see section 5.3). Numerically, the rectangle has advantages similar to those of the strip in alleviating crowding.

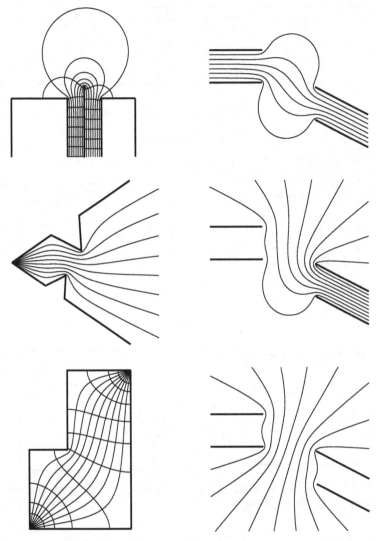

Figure 4.6. More examples of strip maps. In four cases, one set of parallel lines in the strip has been omitted.

As was pointed out in section 2.5, the map to a rectangle requires the specification of four points on the boundary of region P that map to the rectangle's corners. These points make P into a logical or generalized quadrilateral. The locations of these corner images uniquely determine the aspect ratio (conformal modulus) of the target rectangle.

The derivation of the map to polygonally bounded P by the basic formula (4.1) (see p. 48) is rather simple. The Jacobi elliptic function $sn(z \mid m)$

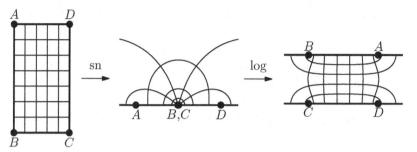

Figure 4.7. Relationships among the rectangle, half-plane, and strip domains. The distortions of the rectangular grid on the left are strong in the middle picture, due to exponential relationships, but the distortions in the strip on the right are relatively minor.

(we usually write simply $\mathrm{sn}(z)$) maps the rectangle to the upper half-plane and the four points $-K + iK'$, $-K$, K, and $K + iK'$ to $-m^{-1/2}$, -1, 1, and $m^{1/2}$, respectively, where m is the **elliptic parameter**.[2] Hence we conclude that

$$f'(z) = C\,\mathrm{sn}'(z) \prod_{k=1}^{n} (\mathrm{sn}(z) - \mathrm{sn}(z_k))^{\alpha_k - 1}. \tag{4.5}$$

The form of (4.5) is not ideal for numerical computation, however. Elliptic functions are expensive to evaluate, and one must explicitly account for the singularity at iK'. Instead, as suggested in [HT90], we find preurtices on the strip and then transplant to the rectangle from there, as shown in Figure 4.7.

Because m is related exponentially to the conformal modulus of the quadrilateral (see (2.12)), points which are spaced algebraically on the rectangle are spaced exponentially on the real line. This fact makes the half-plane numerically unsuitable as an intermediate region. But the additional map $(\log z)/\pi$ maps the rectangle corners to $L + i$, i, 0, and L on the boundary of the strip. Clearly L is linked algebraically to the conformal modulus, so algebraic spacing on the rectangle corresponds to algebraic spacing on the strip.

The situation here is slightly but significantly different from the strip map as described in section 4.2. The conformal modulus, or equivalently the strip parameter L, is an unknown in the problem, and the images of rectangle corners are constrained to lie on two vertical lines, one of which is $\mathrm{Re}\,z = 0$. Altogether this leaves us with $n - 3$ unknown parameters on the strip, and these are determined by $n - 3$ side length conditions exactly as in the standard half-plane and disk maps. Note that the ends of the strip S map to unremarkable points on the boundary of P and play no special role. As usual, the unknowns should

---

[2] The elliptic modulus k is sometimes used instead, the relationship being $m = k^2$.

be transformed so that they are unconstrained (e.g., by taking logarithms of positive differences between prevertices.)

The solution of the parameter problem implicitly determines the correct value of L, and hence the elliptic parameter $m = e^{2\pi L}$, $K(m)$, $K'(m)$, and the conformal modulus $K'/2K$. To compute the map at specific points, one can transplant those points to the strip by the elliptic function and then compute the SC strip map. Some examples are shown in Figures 4.8 and 4.9.

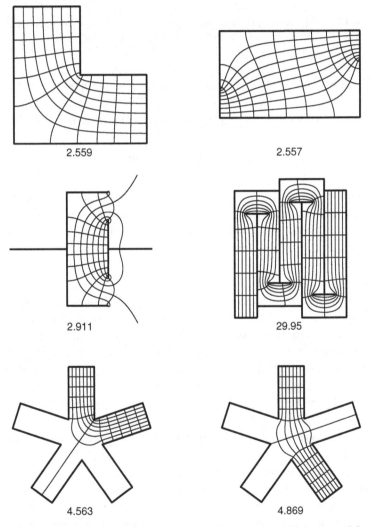

Figure 4.8. Examples of rectangle maps. A regular cartesian grid is mapped from a preimage rectangle to the given region. Underneath each region is its conformal modulus (the aspect ratio of its equivalent rectangle).

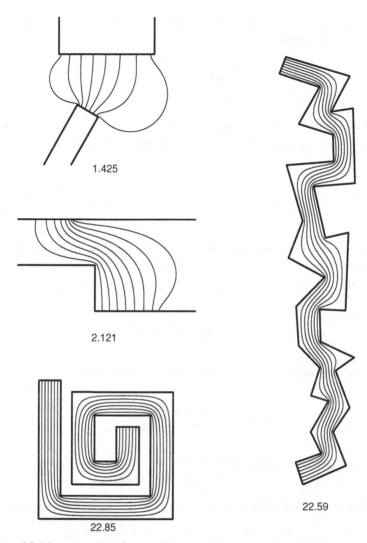

Figure 4.9. More examples of rectangle maps. Only one set of parallel lines from the rectangle is mapped.

4.4 Exterior maps

Suppose now that P is the region exterior to a bounded polygon Γ. In the extended complex plane, P is simply connected, so a conformal map $f(z)$ from the interior of the unit disk to P is guaranteed to exist. To derive this map, we first note that, as one traverses the unit circle in a counterclockwise sense, one must traverse the polygon Γ clockwise so as to keep P on the left. Hence the required jumps in argument are now $(\alpha_k - 1)\pi$ rather than $(1 - \alpha_k)\pi$. From section 4.1 we know that

$$\prod_{k=1}^{n} \left(\frac{z - z_k}{z + 1} \right)^{1-\alpha_k}$$

is a function that has these jumps and is otherwise constant in argument on the unit circle. According to our basic SC principle (4.1), we could set zf' equal to this product:

$$f'(z) = Cz^{-1}(z + 1)^{-2} \prod_{k=1}^{n} (z - z_k)^{1-\alpha_k}.$$

However, this effort would lead to an f that has a logarithmic singularity at the origin. Note that $\arg[(z + 1)^2/z] = 0$ for z on the unit circle, so a geometrically equivalent version is

$$f'(z) = Cz^{-2} \prod_{k=1}^{n} (z - z_k)^{1-\alpha_k}.$$

As usual, we find it more convenient in the disk to write this as

$$f'(z) = Cz^{-2} \prod_{k=1}^{n} \left(1 - \frac{z}{z_k} \right)^{1-\alpha_k}.$$

This form implies that the leading term of f is $-Cz^{-1}$ as $z \to 0$. Hence $f(0) = \infty$ and f is single-valued near the origin. We have found

Schwarz–Christoffel formula for an exterior region from the disk

$$f(z) = A + C \int^{z} \zeta^{-2} \prod_{k=1}^{n} \left(1 - \frac{\zeta}{z_k} \right)^{1-\alpha_k} d\zeta \qquad (4.6)$$

We reiterate that the α_k parameters are related to *interior* angles of the boundary polygon Γ; these are *exterior* to the region P.

Some examples of exterior maps are exhibited in Figures 4.10 and 4.11. One can easily see that

$$z^{-1} = -\frac{1}{C} f(z) + O(1) \quad \text{as } f(z) \to \infty.$$

Figure 4.10. Examples of exterior Schwarz–Christoffel maps. For each region, the curves are images of circles of radius 0.4, 0.5, . . . , 0.9 in the disk.

Consequently, $|C|$ is the **capacity** or transfinite diameter [Ahl78, Hil59, Neh52] of the region P (see also section 5.8).

Computationally, the formula (4.6) is little different from the standard SC formula, although one must take care not to choose an integration path in the disk that passes through or near the singularity at the origin. There is a more significant change, however, in the parameter problem. Because we specified

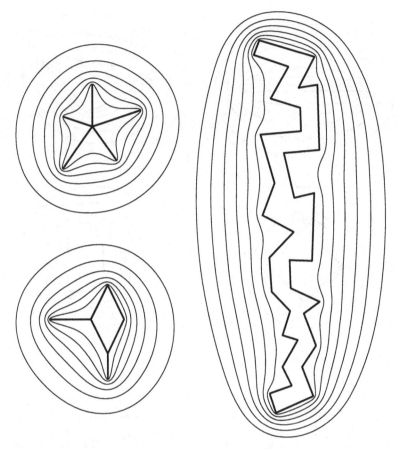

Figure 4.11. More examples of exterior SC maps. The radii of the circles are $0.4, \ldots, 0.9$ on the left, and $0.7, 0.75, \ldots, 0.95$ on the right.

$f(0) = \infty$, only one degree of freedom in f remains (corresponding to a rotation of E). Hence $n - 1$ conditions are required to find the prevertices, yet $n - 3$ conditions determine the polygon uniquely. The remaining two conditions are derived from the single-valuedness of f:

$$0 = \operatorname*{Res}_{z=0} f'(z) = \frac{d}{dz} \left(C \prod_{k=1}^{n} \left(1 - \frac{z}{z_k} \right)^{1-\alpha_k} \right) \Bigg|_{z=0} = C \sum_{k=1}^{n} \frac{\alpha_k - 1}{z_k}. \quad (4.7)$$

This gives the required two extra real conditions.

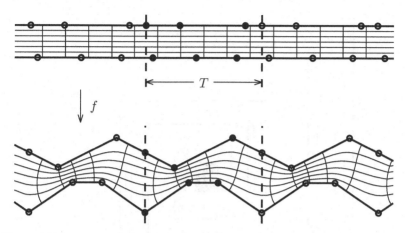

Figure 4.12. Map to a periodic channel from the strip. Only the prevertices of one period (solid dots) and the length of the period need be considered unknowns in the parameter problem. (Dashed lines show periodicity and are not mapped.)

4.5 Periodic regions and fractals

Periodic regions allow the size of the parameter problem to be greatly reduced. Consider the map f from the strip $S = \{0 < \operatorname{Im} z < 1\}$ to a periodic channel such as the one in Figure 4.12. Floryan pointed out that the periodicity of the region implies that the prevertices can be taken to be periodic as well [Flo86, FZ93]. That is, the map can be written as

$$f'(z) = C \prod_{j=-\infty}^{\infty} \prod_{k=1}^{n} \left[\sinh \frac{\pi}{2}(z - z_k - jT) \right]^{\alpha_k - 1},$$

where T is the period of the prevertices. Clearly z_1 can be chosen arbitrarily, leaving $z_2, \ldots, z_n$ and T unknown. As usual, there are $n - 3$ relative side lengths and one complex difference to determine the fundamental geometry; the remaining condition comes from requiring the proper periodicity of the image of one period in z. Computationally, the infinite product can be truncated after just a few terms for any particular z because the hyperbolic sine terms decay exponentially.

Fractals may exhibit analogous regularity. Let f map the strip S to the infinite, self-similar spiral displayed in Figure 4.13. Traveling outward along either side of the spiral, each successive edge is a factor $i\gamma$ times the previous one for some constant $\gamma > 1$. The self-similarity of the sprial implies

$$i\gamma f(S) = f(S). \tag{4.8}$$

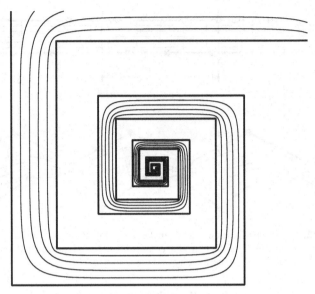

Figure 4.13. Map to a self-similar doubly infinite spiral with similarity factor $\gamma \approx 1.26$. In principle this SC map is completely described by just two unknowns. (The map is approximated here by truncation of an infinite product.)

We assign $f(-\infty) = 0$ and $f(\infty) = \infty$. Only one degree of freedom remains in the map, namely real translation of S. There are infinitely many prevertices on each side of S. If we let z_A and z_B be adjacent prevertices on the bottom of S, we can apply (4.8) to conclude

$$i\gamma f(z) = f(z_B - z_A + z), \qquad \text{for all } z \in S. \qquad (4.9)$$

because both sides of the equation map to the same region and are identical at three points ($\pm\infty$ and z_A). Furthermore, (4.9) is equally valid using the difference between *any* two consecutive prevertices, on either side of S. So this separation must be a constant (i.e., the prevertices are uniformly spaced, with the same spacing on both sides of the strip).

To compute f, we need to know just two quantities: the separation $z_B - z_A$ and the real offset between prevertices on the top and bottom of S. These values are determined by two geometric conditions on $f(S)$: the length ratio of successive sides and the width of one "leg" of the spiral. This is a simple parameter problem. The infinite product in the SC integrand can again be truncated with little error. More complicated fractals (e.g., the exterior of the Koch snowflake) ought to be treatable using similar ideas, in a manner related to multipole methods [GR88]; some steps in this direction have been taken by Banjai

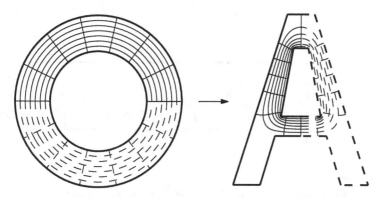

Figure 4.14. Map to a symmetric doubly connected region constructed using a rectangle map with Schwarz reflection.

[Ban00]. As a practical matter, however, one may get good results by truncating the fractal at a finite level of refinement and applying standard SC maps. See [BP93] for an example in the study of drag in fractally bounded channels.

4.6 Reflections and other transformations

So far we have focused on transforming the canonical domain in order to make imposing certain boundary conditions easier or to avoid numerical ill-conditioning. We can also transform the image domain, thereby extending the range of regions that can be mapped.

One important type of such transformations is reflection, which produces symmetric regions. Symmetry is a powerful tool in conformal mapping that appears in many applications (for examples, see [Her82, Hug75]). We have already seen some examples in Figure 2.7. As another example, consider the symmetric, doubly connected polygonal region on the right in Figure 4.14. We can map half of the region to a rectangle, which by suitable scaling and exponentiation becomes half of an annulus. Schwarz reflection allows us to fill in the other half of the annulus and map to the full region.[3] In Figure 4.15, reflection is applied to the rectangle infinitely often to make a strip. The result is a map to a symmetrically regular channel. For an application that makes significant use of symmetry, see section 5.9.

---

[3] Reflection also plays a key role in one derivation of the SC map to general doubly connected regions. See section 4.9.

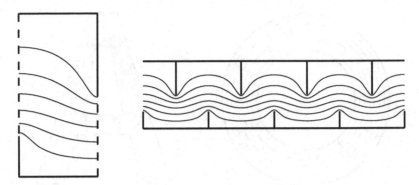

Figure 4.15. Map to a symmetrically periodic channel by repeated reflection of a rectangle map.

Möbius transformations are another interesting way to transform certain polygonal regions. In particular, if any number of circular arcs and straight-line segments meet at a single point, that point can be mapped to infinity and the intersecting sides straightened out. The resulting region is then suitable for an SC map. Some examples of this idea are illustrated in Figure 4.16. These regions are presented as reminders that polygons may hide in unexpected places!

4.7 Riemann surfaces

The classical SC formula is derived under the assumption that the interior angles satisfy

$$\sum_{k=1}^{n}(\alpha_k - 1) = -2.$$

In other words, the exterior turns add up to 2π, or one winding about any interior point. The formula can be extended [Chr70b, Gil49, Goo50, Sch69b], however, to find the map f from the upper half-plane in the case where

$$\sum_{k=1}^{n}(\alpha_k - 1) = -2(b + 1),$$

where b is a natural number. Such a region is no longer planar but a Riemann surface with $b + 1$ sheets.[4] The sheets attach at points $\sigma_1, \ldots, \sigma_b$, which are branch points of the multivalued map f^{-1} to the upper half-plane.

---

[4] Not every self-intersecting polygonal region is a multisheeted surface, however, and the standard SC formula is fine for any single-sheeted region.

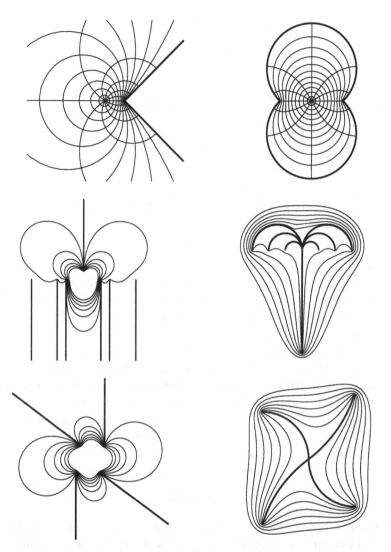

Figure 4.16. Maps using Möbius transformation to straighten circular arcs. The left side shows standard SC maps; on the right are the images of those regions under certain Möbius transformations.

The argument principle implies that f' must have b zeros in the upper half-plane H^+. These points $s_1, \ldots, s_b$ are the preimages of the branch points. We therefore introduce the factors $(z - s_k)$ to f'. To keep them from affecting $\arg f'$ on the real axis, we also must multiply by $(z - \bar{s}_k)$. The map from H^+

to a Riemann surface is thus

Schwarz–Christoffel formula for a Riemann surface

$$f(z) = A + C \int^z \prod_{k=1}^{b} (\zeta - s_k)(\zeta - \bar{s}_k) \prod_{k=1}^{n-1} (\zeta - z_k)^{\alpha_k - 1} d\zeta \qquad (4.10)$$

Because Möbius transformations map inverse points to inverse points, the disk formula can be written as

$$f(z) = A + C \int^z \prod_{k=1}^{b} (\zeta - s_b)(1 - \zeta \bar{s}_b) \prod_{k=1}^{n} \left(1 - \frac{\zeta}{z_k} \right)^{\alpha_k - 1} d\zeta, \qquad (4.11)$$

where $s_1, \ldots, s_b$ are in the unit disk.

The extra $2b$ real unknowns (the roots of f') are determined by requiring that $f(s_k) = \sigma_k$ for $k = 1, \ldots, b$. These unknowns can be transformed to incorporate their natural constraints implicitly. Overall the solution methodology is a fairly minor modification of the usual SC problem. Figure 4.17 illustrates a map to a Riemann surface.

Even though the map to a fully specified, polygonally bounded Riemann surface is readily computed, the reader is urged to think twice before pursuing this technique for an application. In some applications (see sections 5.3 and 5.4), Riemann surfaces arise naturally, but the geometry is discovered, not specified in advance; in other cases (such as the situation discussed in section 5.6 and [ET86]), they arise in some formulations of a problem but can be avoided by more careful ones.

4.8 Gearlike regions

A **gearlike region** is a Jordan region whose boundary segments are all either arcs of circles centered at the origin or segments of rays emanating from the origin. See Figure 4.18. The interior angles of such a region are integer multiples of $\pi/2$. The SC map to such regions was described originally by Goodman [Goo60] (who seems to have first applied the term *gear* in this context) and later by Mason and Jackson [JM87] and by Pearce [Pea91] (who implemented it numerically).

Figure 4.19 shows the logarithms of the regions of Figure 4.18. Each is a Riemann surface whose boundary is made up of horizontal and vertical line segments. One sheet of each surface is shown; dashed lines are to be identified.

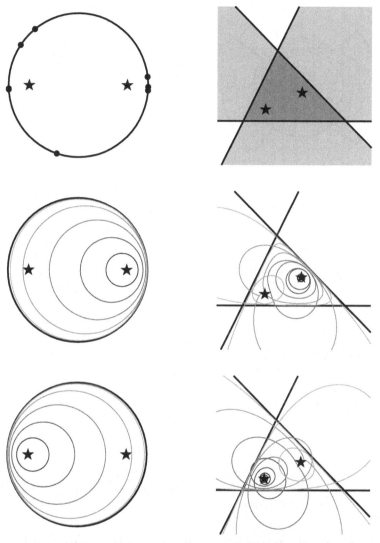

Figure 4.17. Map to a three-sheeted Riemann surface. The unbounded region at top right has three sheets and two branch points (shading indicates winding number). At top left are shown the inverse images of the vertices and the branch points for the SC disk map. The lower two rows show loops around each prebranch point; in the image, each loop circles twice about its central branch point. To enhance clarity, curves have been shaded.

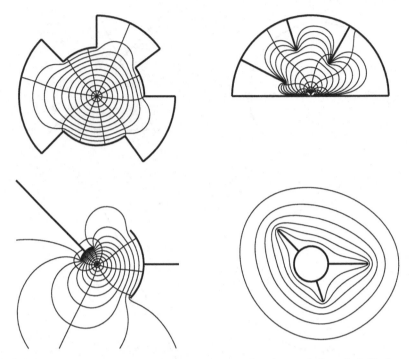

Figure 4.18. Maps to gearlike regions. At the upper right the map is from the half-plane; in the other cases, it is from the disk.

Notice that the appropriate condition on the interior angles is

$$\sum_{k=1}^{n}(\alpha_k - 1) = 0,$$

which corresponds to zero total turn. If the origin is on the boundary of P (as in the upper right of Figure 4.18), only one sheet is needed, and an ordinary SC map to $\log P$ is possible.

To proceed in the general case, we note that the map $\log(f(z))$ from the disk to $\log P$ should have the usual SC boundary behavior. However, if we evaluate this map around a loop enclosing the origin, we should increase its value by $2\pi i$. Hence the integrand for $\log f$ should include a pole with nonzero residue at the origin:

$$(\log f(z))' = Cz^{-1}\prod_{k=1}^{n}\left(1 - \frac{z}{z_k}\right)^{\alpha_k - 1}. \tag{4.12}$$

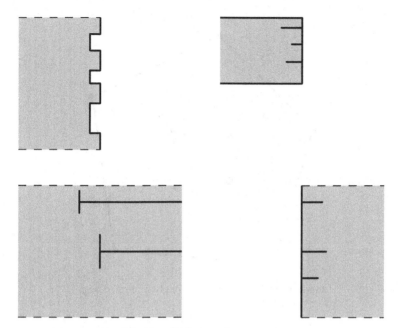

Figure 4.19. Logarithms of the gearlike regions in Figure 4.18.

The residue of the right-hand side at zero is $C \prod(1)^{\alpha_k-1} = C$, so we should choose $C = 1$. (The choice $C = -1$ would be appropriate for an exterior map.) We have obtained

Schwarz–Christoffel formula for a gearlike region

$$f(z) = \exp\left[A \pm \int^z \zeta^{-1} \prod_{k=1}^n \left(1 - \frac{\zeta}{z_k}\right)^{\alpha_k-1} d\zeta \right] \qquad (4.13)$$

We have implicitly imposed the condition $f(0) = 0$, so only one degree of freedom remains in the map (rotation of the disk). Hence there are $n - 1$ unknowns in the parameter problem. Since $\log f$ does not have a free multiplicative constant, we may not rescale the SC image. Moreover, the periodicity of $\log P$ is an additional constraint beyond the $n - 2$ side lengths that determine a regular polygon (see Theorem 3.1). Therefore, $n - 1$ side length conditions are needed to specify $\log P$ correctly.

The map to gearlike regions was described fully by Pearce [Pea91], who also made his software available (see section 3.6).

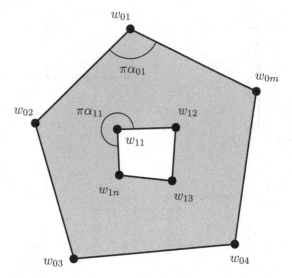

Figure 4.20. Notation for the doubly connected map.

4.9 Doubly connected regions

The SC formula can be extended to doubly connected regions. Suppose P is bounded by an outer polygon, Γ_0, and an inner polygon, Γ_1. The vertices and interior angle parameters of Γ_0 are denoted by $w_{01}, \ldots, w_{0m}$ and $\alpha_{01}, \ldots, \alpha_{0m}$; some of the w_{0k} may be infinite and thus $-2 \le \alpha_{0k} \le 2$ as with the simply connected case. The inner boundary Γ_1 has n vertices, $w_{11}, \ldots, w_{1n}$, which we require to be finite.[5] The angle parameters of Γ_1 are measured in the interior of P and satisfy $0 < \alpha_{1k} \le 2$. See Figure 4.20. From these definitions it is clear that

$$\sum_{k=1}^{m} \alpha_{0k} = m - 2, \qquad \sum_{k=1}^{n} \alpha_{1k} = n + 2.$$

It is well known [Hen86] that there is a unique number $\mu > 0$ such that there exists a conformal map f from the annulus $A_\mu = \{z : \mu < |z| < 1\}$ to P. This map extends continuously to the boundary so that $|z| = \mu$ maps to Γ_1 and $|z| = 1$ maps to Γ_0. The value μ^{-1} is known as the **conformal modulus** of P. The map has one degree of freedom, corresponding to rotation of A_μ. Let $z_{01}, \ldots, z_{0m}$ and $z_{11}, \ldots, z_{1n}$ be the prevertices of Γ_0 and Γ_1 on the outer and

---

[5] The 0 and 1 subscripts are conveniently reminiscent of "outer" and "inner."

inner bounding circles of A_μ, respectively. We have $|z_{0k}| = 1$ and $|z_{1k}| = \mu$. We set $z_{0m} = 1$ to make the map unique.

Application of the basic geometric principle (4.1) is complicated by the presence of two distinct boundary components. We must straighten both components simultaneously in such a way that when a prevertex image is translated to the origin, the argument of the complementary component is constant. Our approach here is a geometric interpretation of a derivation by DeLillo, Elcrat, and Pfaltzgraff [DEP01]. As we shall see, the geometric viewpoint leads quite naturally to the elliptic functions used in practical computations.

Let $z_{0k} \neq 1$ be a vertex on the outer circle $|z| = 1$. The map

$$\frac{1 - (z/z_{0k})}{1 - z} \tag{4.14}$$

straightens out the outer circle while mapping the inner circle $|z| = \mu$ to another circle (see the left-hand side of Figure 4.21). Hence the argument on the inner circle is not constant. We would like to correct for this by multiplying the map by a function that is analytic in A_μ and which takes values conjugate to (4.14) on $|z| = \mu$. Simply conjugating (4.14) would not do, since the result is not analytic. However, reflecting (4.14) across the image of $|z| = \mu$ leaves its value there unchanged. Since reflections are preserved by Möbius transformations, we may

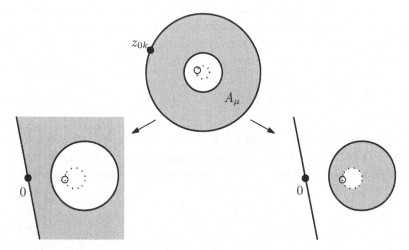

Figure 4.21. First steps toward the doubly connected map. One map appropriately straightens out the outer circle of the annulus but affects the argument on the inner circle (left, (4.14)). By reflecting through the inner circle either before or after mapping, another map can be used to cancel that effect (right, (4.15)).

first reflect in A_μ across $|z| = \mu$ and then apply (4.14). After conjugation, the result is

$$\overline{\left(\frac{1 - \mu^2/(\bar{z}\,z_{0k})}{1 - \mu^2/\bar{z}}\right)} = \frac{1 - \mu^2/(z\,\overline{z_{0k}})}{1 - \mu^2/z} = \frac{1 - \mu^2(z_{0k}/z)}{1 - \mu^2/z}, \qquad (4.15)$$

where in the last step we have used the fact that $|z_{0k}| = 1$. As required, this function is analytic in A_μ. The right-hand side of Figure 4.21 illustrates the action of (4.15) (actually, its conjugate) on the boundary of the annulus.

Although the problem on the inner boundary $|z| = \mu$ has momentarily been resolved, the behavior on the outer boundary $|z| = 1$ has been corrupted. To compensate, we again reflect (this time through $|z| = 1$), apply the most recent map (4.15), and conjugate. The result is

$$\overline{\left(\frac{1 - \mu^2 z_{0k}\,\bar{z}}{1 - \mu^2 \bar{z}}\right)} = \frac{1 - \mu^2(z/z_{0k})}{1 - \mu^2 z}.$$

Now, though, this correction has reintroduced variation of the argument on $|z| = \mu$. If we continue correcting alternately on the two parts of the boundary, we get the infinite product

$$\frac{1 - (z/z_{0k})}{1 - z} \cdot \frac{1 - \mu^2(z_{0k}/z)}{1 - \mu^2/z} \cdot \frac{1 - \mu^2(z/z_{0k})}{1 - \mu^2 z} \cdot \frac{1 - \mu^4(z_{0k}/z)}{1 - \mu^4/z} \cdots . \qquad (4.16)$$

The convergence of the sequence of repeated reflections is illustrated in Figure 4.22: the prevertex z_{0k} is mapped to zero, the outer boundary is mapped to a straight line, and the inner boundary is mapped to a radial segment.

At this point it is useful to introduce the theta function

$$\Theta(z, \mu) = \prod_{j=1}^{\infty}(1 - \mu^{2j-1}z)(1 - \mu^{2j-1}z^{-1}). \qquad (4.17)$$

This function is closely related to the classical Jacobi elliptic theta functions [Hil59, Lan87]. We will henceforth omit the explicit dependence on μ to streamline the notation. With this definition, the product (4.16) simplifies greatly to

$$\frac{\Theta\left(\dfrac{z}{\mu z_{0k}}\right)}{\Theta\left(\dfrac{z}{\mu}\right)}. \qquad (4.18)$$

Convergence is no problem here, though we omit the details. This map introduces a jump in argument at a single prevertex on the outer part of the boundary. If the infinite reflection procedure is applied for a prevertex z_{1k} on the inner

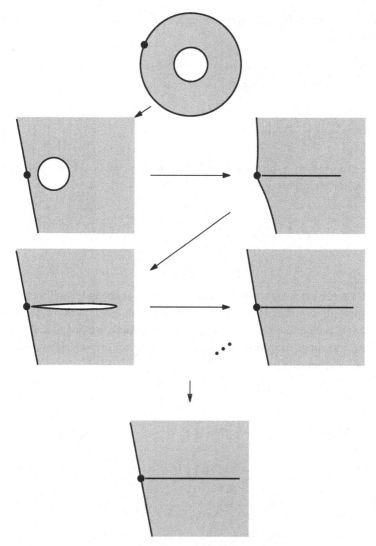

Figure 4.22. Convergence of a doubly connected SC factor. By alternately straightening the outer and inner boundaries of the annulus (left and right), one converges to a map (4.18) which straightens both simultaneously (bottom) and thus has constant argument on one boundary component and constant argument with a single jump on the other. (Although it can be hard to see here, there is always a gap between the dot and the other boundary component.)

boundary, starting the reflections through $|z| = 1$ and using $\overline{z_{1k}} = \mu^2/z_{1k}$, the resulting map is

$$\frac{1 - (z_{1k}/z)}{1 - \mu/z} \cdot \frac{1 - \mu^2(z/z_{1k})}{1 - \mu z} \cdot \frac{1 - \mu^2(z_{1k}/z)}{1 - \mu^3/z} \cdot \frac{1 - \mu^4(z/z_{1k})}{1 - \mu^3 z} \cdots$$

$$= \frac{\Theta\left(\dfrac{\mu z}{z_{1k}}\right)}{\Theta(z)}. \tag{4.19}$$

We are close to a form for f', the derivative of the doubly connected SC map. A first attempt is

$$\prod_{k=1}^{m} \left[\frac{\Theta\left(\dfrac{z}{\mu z_{0k}}\right)}{\Theta\left(\dfrac{z}{\mu}\right)} \right]^{\alpha_{0k}-1} \prod_{k=1}^{n} \left[\frac{\Theta\left(\dfrac{\mu z}{z_{1k}}\right)}{\Theta(z)} \right]^{\alpha_{1k}-1}. \tag{4.20}$$

This form has piecewise constant argument on the inner and outer boundaries, with appropriate jumps. However, the geometry of circles is such that piecewise constant is not quite right for f'. In (4.2), for example, we had to multiply by $(1+z)^{-2}$ to obtain the proper behavior. In fact, $\arg(zf')$ must be kept piecewise constant. Fortunately, we note that if $|u| = 1$,

$$\arg \Theta(u) = \arg \prod |1 - \mu^{2j-2}u|^2 = 0,$$

$$\arg \Theta(u/\mu) = \arg(1 - u) + \arg \prod |1 - \mu^{2j}u|^2 = \tfrac{1}{2}\arg u + \text{const},$$

$$\arg \Theta(\mu u) = \arg(1 - \bar{u}) + \arg \prod |1 - \mu^{2j}u|^2 = -\tfrac{1}{2}\arg u + \text{const}.$$

Hence

$$\arg \left(\left[\Theta\left(\dfrac{z}{\mu}\right) \right]^2 [\Theta(z)]^{-2} \right) = \arg z$$

if $|z| = \mu$ or $|z| = 1$. These results mean that we can remove the denominators in (4.20) and obtain the correct boundary behavior. Finally, we obtain

SC formula for a doubly connected region

$$\boxed{f(z) = A + C \int^z \prod_{k=1}^{m} \left[\Theta\left(\frac{\zeta}{\mu z_{0k}}\right) \right]^{\alpha_{0k}-1} \prod_{k=1}^{n} \left[\Theta\left(\frac{\mu \zeta}{z_{1k}}\right) \right]^{\alpha_{1k}-1} d\zeta} \tag{4.21}$$

This formula is derived rigorously in [DEP01, Hen86, Kom45]. As is pointed out in [DEP01], there are SC singularities at the prevertices and all of their images under reflections through the boundary of the annulus A_μ. Note also that a logarithm maps the annulus to a rectangle, and the repeated reflections just expose the doubly periodic nature of elliptic functions.

In hindsight, we could have derived the strip map (see section 4.2) in the same manner, reflecting alternately across the two sides of the strip. The key identity is

$$(z - z_k)\left(1 - \frac{z - z_k}{2i}\right)\left(1 + \frac{z - z_k}{2i}\right)\left(1 - \frac{z - z_k}{4i}\right)\cdots$$
$$= \frac{2}{\pi}\sinh\left[\frac{\pi}{2}(z - z_k)\right].$$

The resulting factor is exactly the same as (4.3). Here our reflections lead us to a singly periodic function.

A numerical implementation of the doubly connected SC map faces three major issues not present in the simply connected case. These were all noted and effectively dealt with by Däppen [Däp87, Däp88] and Hu [Hu95, Hu98]:

1. The theta function (4.17) must be computed efficiently. It can be expressed as a series, and, if certain identities are used to speed convergence near $\mu = 1$, no more than eight terms are needed to get full double precision.
2. Integration paths must be chosen carefully because a convenient straight-line path often passes outside of A_μ. (Since the integrand is also analytic for $\mu^2 < |z| < \mu$, not every such path is forbidden.) In general one may take a segmented path using arcs of circles and segments of radii.
3. The modulus μ^{-1} must be considered an unknown in the parameter problem. It is convenient to transform the constrained parameter $0 < \mu < 1$ to the unconstrained quantity

$$\frac{1}{\mu_U - \mu} - \frac{1}{\mu - \mu_L},$$

where μ_L and μ_U are bounds for μ. Hu [Hu98] recommended using $\mu_L > 0$ and $\mu_U < 1$ to speed convergence in the solution of the parameter problem.

Examples of doubly connected maps computed using Hu's DSCPACK are shown in Figure 4.23. Following [Däp87, Däp88], we point out that the map represented by Figure 4.22 has another use: upon taking a logarithm, one obtains an infinite channel with a parallel slit. This transformation can be useful as a computational domain for problems of flow with an obstacle.

Figure 4.23. Examples of doubly connected SC maps. Underneath each map is shown the conformal modulus, μ^{-1}. These images were computed using Hu's DSCPACK.

4.10 Circular-arc polygons

A **circular-arc polygon** is a closed curve consisting of finitely many arcs of circles as well as straight-line segments. This extension is a major change in geometry and invalidates our guiding principle (4.1). Hence we revert to an analytic description of the map, using ideas that go back to Schwarz's original paper [Sch69b].

Recall that in the proof of Theorem 1.1 in section 2.2, the key element was the invariance of f''/f' under linear (affine) transformation—the kind of transformation that results from two reflections. Similarly, two reflections through circular arcs always produce a Möbius transformation, so the key to an SC map in this case will be an expression in f and its derivatives that is invariant under Möbius transformations. Such an expression is the **Schwarzian**,

$$\{f, z\} = \left(\frac{f''(z)}{f'(z)}\right)' - \frac{1}{2}\left(\frac{f''(z)}{f'(z)}\right)^2.$$

As before, at a prevertex of the circular-arc polygon, the map has the form $(z - z_k)^{\alpha_k}\psi(z)$ for analytic ψ. (This always creates a wedge with straight sides, but those can be mapped to circular arcs by Möbius transformation while leaving the Schwarzian unchanged.) Taking the Schwarzian, we obtain

$$\frac{1}{2}\frac{1 - \alpha_k^2}{(z - z_k)^2} + \frac{\gamma_k}{z - z_k},$$

where, as usual, $\pi\alpha_k$ is the interior angle, and

$$\gamma_k = \frac{1 - \alpha_k^2}{\alpha_k}\frac{\psi'(z_k)}{\psi(z_k)}.$$

These are the only singularities in $\{f, z\}$, and Liouville's theorem implies

SC differential equation for a circular-arc polygon

$$\boxed{\{f, z\} = \frac{1}{2}\sum_{k=1}^{n}\frac{1 - \alpha_k^2}{(z - z_k)^2} + \sum_{k=1}^{n}\frac{\gamma_k}{z - z_k}} \qquad (4.22)$$

For all the details of the derivation, see [AF97, Neh52, Sch69b]. To define the equation for maps from the unit disk, write (4.22) as $\{f, z\} = S(z)$ and let $\phi(z) = i(1 - z)/(1 + z)$ be a map of the disk to H^+. Then

$$\{f, z\} = S(\phi(z))[\phi'(z)]^2$$

is the disk formula [BG87].

Equation (4.22) is a third-order differential equation for the mapping function f. The unknown parameters are the prevertices $z_1, \ldots, z_n$ and the new values

$\gamma_1, \ldots, \gamma_n$. They satisfy the following constraints:

$$\sum_{k=1}^{n} \gamma_k = 0,$$

$$\sum_{k=1}^{n} [2\gamma_k z_k + (1 - \alpha_k)^2] = 0,$$

$$\sum_{k=1}^{n} \left[\gamma_k z_k^2 + \left(1 - \alpha_k^2\right) z_k \right] = 0,$$

as well as the usual $z_1 < \cdots < z_n$. Note that there is no longer a constraint on $\sum \alpha_k$, as the geometry is more flexible.

As always with simply connected regions, there are three degrees of freedom in the map. To specify the mapping function uniquely, we can give f, f', and f'' at a point. (Modifying these conditions corresponds to applying a Möbius transformation to the image.) Equivalently, we could specify the locations of three prevertices instead. Thus a polygon with three arcs (a circular triangle) requires no parameter problem to be solved, and in fact the map can be expressed in terms of hypergeometric functions [GR94]. Such maps, and their inverses, have applications in the study of relativity and integrable systems; for a thorough discussion, see [AF97, Example 5.8.2].

In the general case, the parameters $\gamma_1, \ldots, \gamma_n$ are a serious new obstacle to numerical implementation. The core of the difficulty appears to be that they cause the nonlinear system of equations describing the parameter problem to be highly ill conditioned. Another complication arises when an angle α_k is equal to 0, 1, or 2, signaling that two arcs meet tangentially. In that case the preceeding discussion and formula need to be modified because f has a logarithmic singularity at z_k.

Two attempts, detailed in [BG87] and [How93], have been made to implement the circular-arc formula numerically in a general setting. The method of Bjørstad and Grosse uses the disk formulation and performs integration from the origin to the midpoints of prevertex arcs. The vertices are found as the intersections of arcs (assuming noninteger values of α). This approach avoids the singularities in the differential equation but breaks down in the presence of crowding. Their software is available from Netlib (see section 3.6). Howell improved the performance of the numerical integration by integrating along the boundary of the half-plane, thereby using only real arithmetic, and by transforming the singularities away. (A different form of the equation must be used if the integration extends into the half-plane.) Howell also transformed the γ-parameters to improve conditioning somewhat, but ill-conditioning persists, and the solution of the parameter problem remains quite difficult in general.

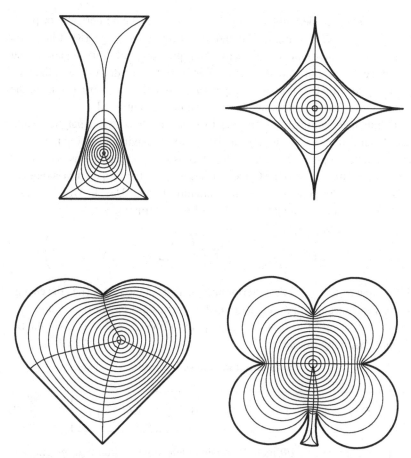

Figure 4.24. Maps to circular-arc polygons. Data for these plots were produced by Louis Howell using unreleased FORTRAN software.

However, one can find maps to many interesting regions in practice, as illustrated in Figure 4.24.

Computational use of the circular-arc map is considerably more difficult than in the straight-side case, and neither of the two implementations descibed previously is as robust or powerful as other SC software. Perhaps a new idea, such as a geometric interpretation of the Schwarzian and the γ_k parameters, is needed before fully satisfactory circular-arc polygon software can be written.

4.11 Curved boundaries

Although the essence of classical Schwarz–Christoffel mapping is to account for corners, it is possible to adapt the idea to smooth, curved boundaries. For example, since corners induce singularities that may be numerically or physically undesirable (see Figure 2.8 and the discussion on page 18), it has been suggested that the standard SC factors be replaced by arbitrary functions that introduce a rapid but smooth change of argument in the derivative of the map [Hen48, Hen86]. This process tacitly assumes that the geometric details of the rounding may not be important and so need not be controlled.

Of much greater interest is the map to a specified, piecewise analytic boundary. A general curved boundary can be viewed as having a turning tangent at every point, not just at finitely many points as in a polygon. It has been noted many times that the basic SC idea of introducing the proper turn along the boundary can be generalized to this situation [CS92, Dav79, DE93, Lea15, Woo61]. We first rewrite the classical half-plane formula:

$$f'(z) = C \prod_k (z - z_k)^{\alpha_k - 1} = C \exp\left(-\sum_k (1 - \alpha_k) \log(z - z_k)\right).$$

Recall that $\pi(1 - \alpha_k)$ is the exterior turning angle at the vertex (i.e., the change in the direction angle of the tangent). In the general case, let $\theta(x)$ be the direction angle of the tangent to Γ at the image point of the real value x. Then the SC formula naturally extends to

SC formula for curved boundary

$$\boxed{f(z) = A + C \int^z \exp\left(-\frac{1}{\pi} \int_{-\infty}^{\infty} \log(\zeta - x)\, d\theta(x)\right) d\zeta} \qquad (4.23)$$

At a true corner in Γ, $d\theta$ includes a Dirac delta contribution that integrates to recover a standard SC term.

The primary difficulty in applying (4.23) is determining the unknown function $\theta(x)$. Indeed, determining the **boundary correspondence** is the key step in most numerical methods for conformal mapping—the classical SC prevertex parameter problem being a special case. Davis [Dav79] used a piecewise quadratic representation of $\theta(x)$ and midpoint quadrature in a boundary element method to find $f'(z)$. (Upon integration this form becomes piecewise cubic, allowing points and slopes to be matched on the boundary at the breakpoints.) Finding the boundary correspondence in this case reduces to finding the preimages of the breakpoints on Γ. The ability to build in accurate treatment of the corners of a piecewise smooth boundary is attractive, but other integral equation methods for conformal mapping are probably preferable and have certainly received more attention [DeL87, Gai64, Hou90, Kyt98, Tre86, Weg86].

5

Applications

Conformal mapping in general, and Schwarz–Christoffel mapping in particular, are fascinating and beautiful subjects in their own rights. Nevertheless, the history of conformal mapping is driven largely by applications, so it is appropriate to consider when and how SC mapping can be used in practical problems.

It is not our intent in this final chapter to recount every instance in which Schwarz–Christoffel mapping has been brought to bear. Rather, after a brief look at a few areas full of such examples, we describe some situations in which SC ideas can be applied in ways that are computational and perhaps not transparent. The most famous application of conformal mapping is to Laplace's equation, and we devote three sections to it. Beyond this it is clear that Schwarz–Christoffel mapping has a small but important niche in applied mathematics and science.

In applications it is common to pose a physical problem in the z-plane, which maps to a canonical region in the w-plane. This convention runs counter to our discussion in the earlier chapters, in which w was the plane of the polygon. In the following sections we attempt to be consistent with established applications literature where appropriate.

5.1 Why use Schwarz–Christoffel maps?

Schwarz–Christoffel mapping is an incomparably effective tool for a very specific sort of problem. The most natural and satisfying application is the solution of Laplace's equation in the plane with piecewise constant (and in the case of derivative conditions, homogeneous) boundary conditions. All elements of this problem are important to varying degrees—a forcing (Poisson) term or a different type of boundary condition makes a big, if not always fatal, difference, and moving to three dimensions removes conformal mapping from consideration.

75

Even so, Laplace's equation is one of the most fundamental in physics, and two-dimensional insights are still valuable for many phenomena. Less-conspicuous applications of conformal mapping are also plentiful. For more on the role of conformal mapping in applications, see [Hen86, SL91].

For problems in which Schwarz–Christoffel mapping is appropriate, it is very powerful. Many researchers have used SC maps of simple geometries to get fully analytical solutions to applied problems. Numerical SC mapping is typically fast and accurate enough that it can also be seen as providing an explicit, though numerical, solution to such problems. In this regard SC mapping is quite different from conformal mapping methods based on, say, integral equations, where the necessary effort and expected accuracy of the conformal mapping may well be comparable to those associated with solving Laplace's equation on the original domain.

One field with an abundance of relevant phenomena is electrical engineering. Standard problems of determining electrical resistance, capacitance, or electrical or magnetic potential can be solved by mapping to a rectangle (section 4.3), or, more generally, by the techniques of sections 5.2–5.4. When mapping is used as part of a design or inverse iteration, the iteration can often be moved "inside" the mapping process (see section 5.5). For classical electromagnetic applications, see [BL63, Gui50, Hal67, OL96, Pal37]. SC maps have been applied to microwave waveguides [Cos87, Cos01, GB+01], integrated circuits [Cha89, Cha92, HD86, KO89, PS99], magnetoresistive disk drive heads [HLD99], magnetic motors [CCBS00], automatic control [CST95], resistor trimming [Nic97, Tre84], crack detection [EH96, EIN95], Van der Pauw resistance measurement [Ver83], and the Hall effect [TW86, Ver82].

Another of the major classical sources of conformal mapping applications is fluid mechanics. Applications in potential flow go back to the nineteenth century and are described in numerous books such as [Lam45]. The special case of potential flows involving jets, wakes, and cavities has an equally long literature going back to the 1860s, just like the Schwarz–Christoffel transformation itself; see section 5.6. Because real flows usually involve vorticity and separation or three-dimensional effects, conformal mapping is used less often than it used to be to provide explicit solutions of flow problems. However, it remains important as a tool for imposing no-flow boundary conditions in more complicated fluid mechanics calculations via mapping of a complicated 2D flow domain to a half-plane, where the method of images can be applied [GC87, Mil96, PH95].

Mesh generation was the primary motivation for some investigators of numerical SC mapping [ADHE82, Dav79, Hoe86, Ive82, Sku66] and has been used successfully in applications [BEMR94, CF98, MRB94, SD85].

Historically, though, mesh generation has played a larger role in the development of SC mapping than vice versa. See section 5.7.

5.2 Piecewise-constant boundary conditions

The simplest case of Laplace's equation occurs when the solution takes a constant value everywhere on the boundary. This problem is trivial, of course, unless some point is singled out to have a different behavior, as in a **Green's function**. Given a region D bounded by Γ and a source point $z_0 \in D$, the Green's function $g(z)$ is defined by the conditions

$$\Delta g(z) = 0 \qquad \text{for } z \in D\backslash\{z_0\}, \tag{5.1a}$$

$$g(z) \to 0 \qquad \text{for } z \to \Gamma, \tag{5.1b}$$

$$g(z) \sim -\log|z - z_0| \qquad \text{for } z \to z_0. \tag{5.1c}$$

One interpretation of the Green's function is as the electrical potential in a vacuum due to a point charge at z_0 and a perfectly conducting surface Γ (both extended infinitely through the third dimension). In the disk map, z_0 is the conformal center. Thus, Figures 4.1 and 4.2 show equipotentials and flow lines for a point charge, as the reader's physical intuition probably assumed. For the exterior map, $z_0 = \infty$ is the natural choice (in which case (5.1c) reads $g(z) \sim \log|z|$), and Figures 4.10 and 4.11 can be interpreted accordingly.

The strip map of section 4.2 solves a slightly different problem: Laplace's equation with *two* distinct Dirichlet boundary values. The procedure is illustrated in Figure 5.1. The entire boundary is decomposed into two components corresponding to the boundary values ϕ_1 and ϕ_2. Each component is the image

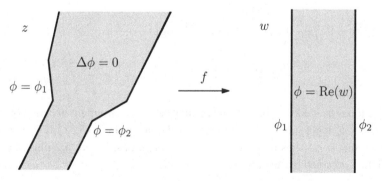

Figure 5.1. Solving Laplace's equation with two Dirichlet boundary values by mapping to a strip. See also Figures 4.5 and 4.6.

under $z = f^{-1}(w)$ of one side of the strip. The strip can be rotated, translated, and scaled as necessary so that its sides are vertical and line up with real parts equal to ϕ_1 and ϕ_2. In the w-plane, the function $\phi = \text{Re } w$ is clearly harmonic and satisfies the boundary conditions. In terms of the original variable, we have $\phi(z) = \text{Re} f(z)$. Hence each of the pictures in Figures 4.5 and 4.6 can be viewed as a solution to a potential problem with two boundary values. The images of the strip "ends" are the points where ϕ change value, and they can be selected independently of the geometry (see the right-hand column of Figure 4.6). A map to a doubly connected region from an annulus can also be interpreted as the solution to a two-value Laplace problem. In this case the appropriate potential in the w-plane of the annulus is $\log |w|$ (see Figure 4.23 for examples).

In the simply connected case, the procedure is easily generalized to $\kappa > 2$ distinct boundary values. First we map a polygonal region onto the half-plane H^+. (Because this straightforward step is implied throughout this and the next two sections, we will refer to the half-plane variable as z.) This step gives us points $-\infty < x_1 < x_2 < \cdots < x_{\kappa-1} < \infty$, as well as ∞ itself, at which the boundary conditions may change.[1] Given κ values $\phi_1, \ldots, \phi_\kappa$, we seek a real function $\phi(z)$, harmonic in H^+, such that

$$\phi(x) = \begin{cases} \phi_1, & -\infty < x < x_1 \\ \phi_2, & x_1 < x < x_2 \\ \vdots \\ \phi_{\kappa-1}, & x_{\kappa-2} < x < x_{\kappa-1} \\ \phi_\kappa, & x_{\kappa-1} < x < \infty, \end{cases} \tag{5.2}$$

The solution of this boundary value problem is

$$\phi(z) = \text{Re}\left[-\frac{i}{\pi} \left(\phi_1 \log(z - x_1) + \phi_2 \log\left(\frac{z - x_2}{z - x_1}\right) \right. \right.$$
$$\left. \left. + \cdots + \phi_{\kappa-1} \log\left(\frac{z - x_{\kappa-1}}{z - x_{\kappa-2}}\right) - \phi_\kappa \log(x_{\kappa-1} - z) \right) \right]. \tag{5.3}$$

(Caution is needed with branch cuts; it may be more convenient in computation to use $+\phi_\kappa \log[1/(x_{\kappa-1} - z)]$ in place of the last term in (5.3).) It is easy to verify that ϕ satisfies the stated boundary conditions. Figure 5.2 illustrates a solution obtained by this formula.

---

[1] Throughout this discussion and the next two sections, we ignore the potentially serious computational issue of crowding. See sections 2.6 and 3.4.

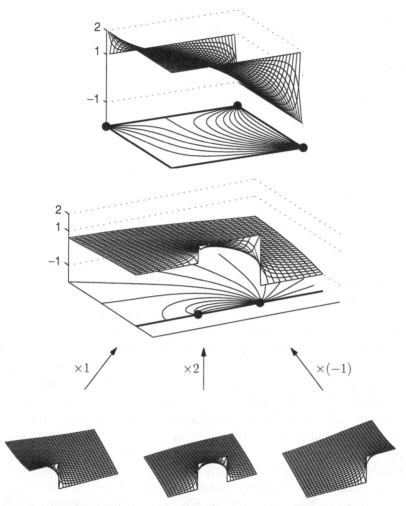

Figure 5.2. Components of a harmonic function with piecewise-constant boundary values. At top, the problem is posed on a square with three distinct boundary values. In the middle, the problem is transplanted to the half-plane. The bottom row shows how each boundary component contributes linearly to the final solution according to (5.3).

We have found $\phi(z)$ as the real part of an analytic map $f(z)$. Differentiating, we obtain

$$\frac{df}{dz} = -\frac{i}{\pi}\left[\frac{\phi_1}{z-x_1} + \frac{\phi_2(x_2-x_1)}{(z-x_1)(z-x_2)} + \cdots - \frac{\phi_\kappa}{z-x_{\kappa-1}}\right]$$

$$= i\frac{p(z)}{(z-x_1)(z-x_2)\cdots(z-x_{\kappa-1})}, \tag{5.4}$$

where $p(z)$ is a real polynomial of degree at most $\kappa - 2$. Clearly, f is a Schwarz–Christoffel map! We summarize this observation as follows.

Theorem 5.1. *Given* $m > 1$, *real values* $\phi_1, \ldots, \phi_\kappa$, *and* $-\infty < x_1 < x_2 < \cdots < x_{\kappa-1} < \infty$, *there exists a unique real polynomial* p *of degree at most* $\kappa - 2$ *such that*

$$\phi(z) = \phi_1 + \text{Re}\left[i \int_{x_1}^{z} p(\zeta) \prod_{k=1}^{\kappa-1} (\zeta - x_k)^{-1} d\zeta \right] \tag{5.5}$$

is the unique harmonic function in H^+ *satisfying* (5.2).

An examination of (5.4) reveals that if x_j is a root of $p(z)$, then $\phi_j = \phi_{j+1}$. In other words, a common factor in the fraction of (5.4) occurs if and only if two adjacent boundary values are equal; hence, no special treatment is needed between them. Similarly, the leading term of p is $(\phi_1 - \phi_\kappa)z^{\kappa-2}$. If $\phi_1 = \phi_\kappa$, then no special behavior is needed at infinity, and the map will reflect this.

Suppose that we exclude these degenerate situations. Then $x_1, \ldots, x_{\kappa-1}$ are mapped by f to infinity in such a way that adjacent edges are tangent there. The real roots of $p(z)$ are mapped to the tips of slits. If p has $\kappa - 2$ real roots, then the exponents in the integrand of (5.5) sum to $\kappa - 2 - (\kappa - 1) = -1$, and thus f maps ∞ to ∞ with tangent sides as well. If $\kappa \geq 4$, complex conjugate roots are possible. Each pair of complex roots for p removes two slits, the sum of exponents decreases by 2, and the image under f is a multiple-sheeted Riemann surface (see section 4.7). In every case, the image under f has only vertical sides because those correspond to piecewise-constant Dirichlet conditions. Some possibilities for four distinct boundary values ($\kappa = 4$) are shown in Figure 5.3 for an L-shaped region. Examples for other regions are shown in Figure 5.4.

In summary, here is the solution procedure for the piecewise-constant Dirichlet problem for Laplace's equation in a polygon:

1. Find an SC map to transplant the polygon to the upper half-plane, solving the usual nonlinear parameter problem.
2. Find a second SC map (5.5) whose integrand involves the "jump points" on the boundary as found in step 1, times a polynomial term that can be deduced from the same data.
3. Take the real part of the composite map.

Of course, in practice it is much easier to use (5.3) than the seemingly circuitous step 2. But in the next two sections we shall see that step 2 can be modified

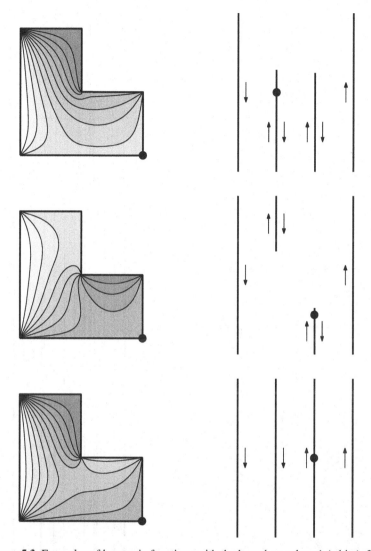

Figure 5.3. Examples of harmonic functions with the boundary values 1 (white), 2, 3, 4 (medium gray) in different orderings on an L-shaped domain. Each function is the real part of the map to the region at the right. Level curves are shown on the left, and the image of one point is shown for reference. Arrows on the right correspond to a counterclockwise traversal around the L. The region at bottom right is a two-sheeted surface.

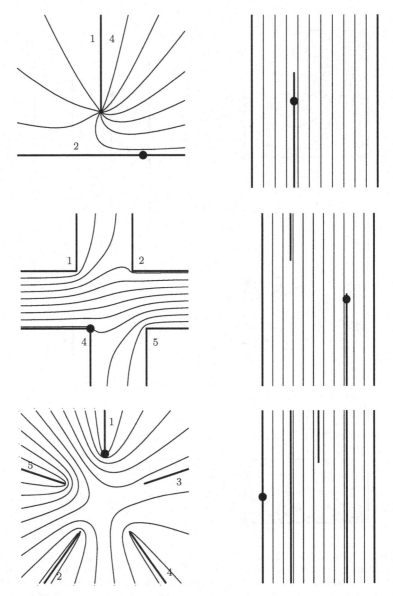

Figure 5.4. More examples of harmonic functions with piecewise-constant Dirichlet boundary data. Level curves are shown in both the original (left) and target (right) geometries. Boundary values are indicated in the left column.

slightly to solve Laplace problems that have derivative boundary conditions on some sides.

5.3 Alternating Dirichlet and Neumann conditions

In the previous section we showed that Laplace's equation with piecewise-constant Dirichlet boundary conditions can be solved by composing two SC maps, one standard and one whose integrand is a rational function. In this section we solve the related problem wherein Dirichlet values are interleaved with homogeneous Neumann (normal derivative) conditions.[2] Techniques closely related to those in this section have appeared numerous times in the literature; for example, see [ET99, LL83, Ver83, Wid69].

As in the preceding section, we assume that a problem on a polygon is first transplanted to the half-plane, whose variable we call z. The real axis is therefore divided into intervals (determined by the original geometry) on which we alternate constant Dirichlet and homogeneous Neumann conditions. In section 5.2 we mapped the real axis to a polygon with all vertical sides because taking the real part led naturally to Dirichlet conditions. Here the appropriate image geometry is a polygon with alternating vertical and horizontal sides, corresponding to Dirichlet and homogeneous Neumann conditions, respectively. We have seen such behavior before, from rectangle maps (section 4.3). Every plot in Figures 4.8 and 4.9, for instance, can be interpreted as a solution to a potential problem with D/N/D/N boundary conditions. The Dirichlet values, as with the strip, can be adjusted arbitrarily by real scaling and translation.

Unlike the situation with the strip, where two adjacent Dirichlet values cause a logarithmic singularity, the rectangle leads to a continuous, regular solution. As Figure 5.5 suggests, however, infinitely many unbounded solutions are possible. Indeed, if one tries to draw pictures for the simpler case of just one constant Dirichlet interval, one soon (and correctly) concludes that all valid pictures lead to unbounded solutions. Since any point in a Neumann interval may map to infinity, there are uncountably many unbounded solutions for every problem.

In some applications, unbounded solutions might be appropriate (for one example, see section 5.9). But if we exclude them, the standard theory of harmonic functions [Joh82] assures us that the solution of the boundary-value problem is unique. In the case of one Dirichlet interval, the solution is constant. In a problem with two or more intervals, just one map corresponds to the bounded solution. Hence the rectangle completely solves the bounded D/N/D/N problem.

---

[2] Inhomogeneous derivative conditions do not transplant trivially and therefore are much less amenable to conformal mapping.

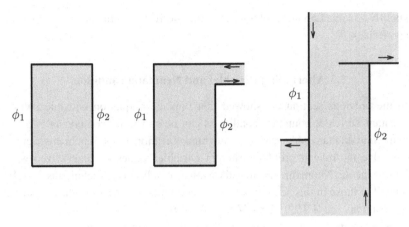

Figure 5.5. Solutions to the D/N/D/N Laplace problem. As the real parts of these images suggest, infinitely many solutions are possible, but the only bounded solution is provided by the map to a rectangle.

We now generalize the rectangle to any number $\kappa > 1$ of Dirichlet values. Figure 5.6 suggests how to find a harmonic function ϕ with $\kappa = 3$. We look for an SC map from the half-plane H^+ such that the three intervals on which the solution is constant are each mapped to vertical segments, while the rest of the real axis maps to horizontal sides. The real part of any such map automatically satisfies the proper *types* of boundary conditions, but the vertical sides must also have prescribed locations. The usual scaling and translation can assign only two of them. To get the flexibility needed for the remaining value, we introduce a new factor in the SC integrand corresponding to a slit, which does not change the angle a side makes with the positive real axis.

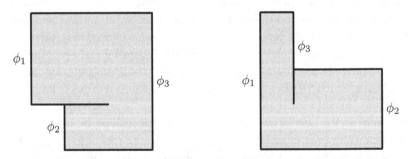

Figure 5.6. Solutions of alternating Dirichlet-Neumann problems with three prescribed Dirichlet values. The real part of each map provides the solution. A slit, whose location is determined by the original geometry and the three prescribed values, is needed to keep the solution bounded.

The endpoints of the Dirichlet intervals are given (perhaps deduced from an original polygonal geometry), and the values to be taken by ϕ on them are prescribed. Thus the only unknowns are the location of the slit and the global scaling and translation. What makes the problem remarkable—and this is not geometrically obvious in Figure 5.6—is that the associated parameter problem is *linear*, and this fact is true regardless of the number of Dirichlet values being assigned.

Theorem 5.2. *Let $\kappa > 1$, real values $\phi_1, \ldots, \phi_\kappa$, and disjoint real intervals*

$$
\begin{aligned}
I_1 &= (-\infty, x_1) \\
I_2 &= (u_2, x_2) \\
&\vdots \\
I_\kappa &= (u_\kappa, x_\kappa), \quad x_\kappa < \infty
\end{aligned}
$$

be given. Then there is a unique real polynomial $p(z)$ of degree no greater than $\kappa - 2$ such that

$$
\phi(z) = \phi_1 + \text{Re}\left[\int_{x_1}^{z} p(\zeta)(\zeta - x_1)^{-1/2} \prod_{j=2}^{\kappa} (\zeta - u_j)^{-1/2}(\zeta - x_j)^{-1/2}\, d\zeta \right]
\tag{5.6}
$$

(where the principal branch of the square root is understood) is the unique bounded harmonic function in the upper half-plane such that $\phi(x) = \phi_j$ if $x \in I_j$ and $\partial\phi/\partial n = 0$ for $x \notin \cup_j I_j$.

Proof. Without any loss of generality, we assume that the intervals are given in order, so that $x_j < u_k$ if $j < k$. Define

$$
g(z) = (z - x_1)^{-1/2} \prod_{j=2}^{\kappa} (z - u_j)^{-1/2}(z - x_j)^{-1/2},
\tag{5.7}
$$

and consider the $(\kappa - 1) \times (\kappa - 1)$ matrix

$$
M = \begin{bmatrix}
\int_{x_1}^{u_2} g(\zeta)\, d\zeta & \int_{x_1}^{u_2} \zeta g(\zeta)\, d\zeta & \cdots & \int_{x_1}^{u_2} \zeta^{\kappa-2} g(\zeta)\, d\zeta \\
\int_{x_2}^{u_3} g(\zeta)\, d\zeta & \int_{x_2}^{u_3} \zeta g(\zeta)\, d\zeta & \cdots & \int_{x_2}^{u_3} \zeta^{\kappa-2} g(\zeta)\, d\zeta \\
\vdots & \vdots & & \vdots \\
\int_{x_{\kappa-1}}^{u_\kappa} g(\zeta)\, d\zeta & \int_{x_{\kappa-1}}^{u_\kappa} \zeta g(\zeta)\, d\zeta & \cdots & \int_{x_{\kappa-1}}^{u_\kappa} \zeta^{\kappa-2} g(\zeta)\, d\zeta
\end{bmatrix}.
\tag{5.8}
$$

We observe that M is real because the integrands are all real over the intervals chosen (there is always an even number of imaginary terms in g). Suppose that M is singular. Then there exists a real nonzero polynomial q of degree no greater than $\kappa - 2$ such that the SC map

$$f(z) = \int_{x_1}^{z} q(\zeta) g(\zeta) \, d\zeta$$

satisfies

$$f(u_2) - f(x_1) = 0, \quad f(u_3) - f(x_2) = 0, \ldots, f(u_\kappa) - f(x_{\kappa-1}) = 0.$$

By the nature of the integrand for f, this situation can happen only if q has one root inside each of the Neumann intervals (x_1, u_2), (x_2, u_3), $\ldots$, $(x_{\kappa-1}, u_\kappa)$. This would imply that q has at least $\kappa - 1$ roots, which is impossible. Hence M is nonsingular.

We now solve the linear system

$$M \begin{bmatrix} b_0 \\ b_1 \\ \vdots \\ b_{\kappa-2} \end{bmatrix} = \begin{bmatrix} \phi_2 - \phi_1 \\ \phi_3 - \phi_2 \\ \vdots \\ \phi_\kappa - \phi_{\kappa-1} \end{bmatrix} \tag{5.9}$$

and write $p(z) = b_0 + b_1 z + \cdots + b_{\kappa-2} z^{\kappa-2}$. Define ϕ by (5.6). As the real part of an analytic function (an SC map) in the upper half-plane, ϕ is harmonic. The exponents of the factors of $p(\zeta) g(\zeta)$ imply that ϕ is bounded. For real ζ, $p(\zeta)$ is real, and $g(\zeta)$ is imaginary if ζ is in some I_j and real otherwise. These facts and the imposition of (5.9) guarantee satisfaction of the boundary conditions. Finally, the uniqueness of SC mapping ensures the uniqueness of ϕ, for otherwise we could use the harmonic conjugate of ϕ to construct a different map to the same region.   □

Apart from the new linear parameter problem, there is much similarity between Theorem 5.1 and Theorem 5.2. The real roots of the polynomial p correspond to slits in the image, unless they coincide with an interval endpoint to make a "left turn" into a "right turn." Complex roots, which can appear in conjugate pairs only for $\kappa \geq 4$, decrease the sum of the SC exponents and lead to a Riemann surface, as described in sections 4.7 and 5.2. Figure 5.7 illustrates this possibility for the minimal value $\kappa = 4$. In terms of the original boundary-value problem, one simply concludes that ϕ assumes some values more than once in the region.

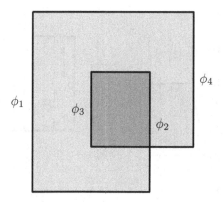

Figure 5.7. A Riemann surface in the four-value D/N problem. If roots of the parameter polynomial in Theorem 5.2 are complex, parts of the plane will be covered more than once.

Indeed, we can recover Theorem 5.1 if we let $x_k \to u_{k+1}$ for $k < \kappa$, and $x_\kappa \to \infty$. The SC product mostly telescopes to give exponents of -1 as in (5.5) (geometrically, the two corners defining a horizontal side retreat to infinity), while the term $(z - x_\kappa)^{-1/2}$ is imaginary on the whole real axis and becomes the extra factor of i in that formula. Thus, while we were able to construct the polynomial p in section 5.2 directly from the problem data, we could also view it as arising from a linear parameter problem.

Linear parameter problems have appeared in diverse places in the literature— for example, in the Van der Pauw resistance measurement configuration [Ver83] and in the zero-dispersion limit of the Korteweg–de Vries equation [LL83]. The idea of Theorem 5.2 is also similar to the "rectified polygon" introduced in section 3.5. There, however, we had neither slits nor a linear parameter problem because we were not interested in assigning particular Dirichlet values.

Figure 5.8 displays some further examples of solutions obtained by Theorem 5.2, for an L-shaped region. Figure 5.9 illustrates examples on a few other regions.

5.4 Oblique derivative boundary conditions

As described in detail in section 5.3, an SC map can be applied elegantly to determine a harmonic function ϕ (i.e., a solution of $\Delta\phi = 0$) that satisfies alternating constant Dirichlet and homogeneous Neumann conditions on the sides of a polygon P. To solve such a problem, we construct (using two SC maps, one of which has a linear parameter problem) a conformal map of P

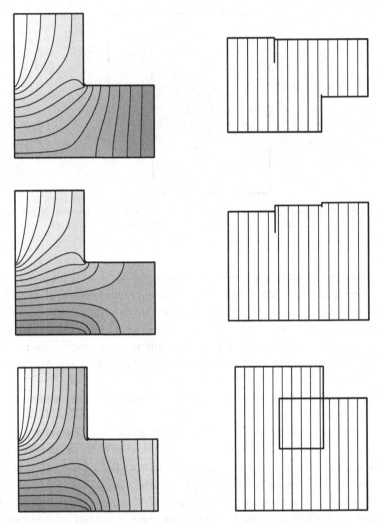

Figure 5.8. Examples of alternating D/N solutions on an L-shaped region. Every problem domain on the left has the same pattern of Dirichlet and Neumann assignments, and the same four Dirichlet values 1, 2, 3, 4, but the order of these values is different in each case. The right column shows the target regions obtained from Theorem 5.2; the solutions are the real parts of these maps. Level curves of each solution are drawn, and on the left these are shaded from 1 (white) to 4 (medium gray).

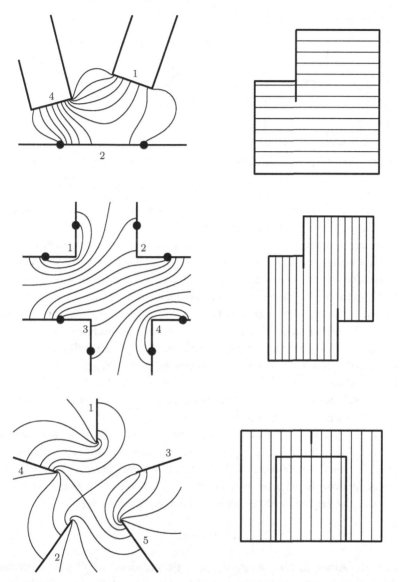

Figure 5.9. Further examples of alternating D/N solutions, on unbounded regions, for $\kappa = 3, 4, 5$ (top to bottom). The assigned Dirichlet values are shown in each case. At the top we have plotted flow lines rather than equipotentials. At the bottom, note that two curves meet at the preimage of a branch point.

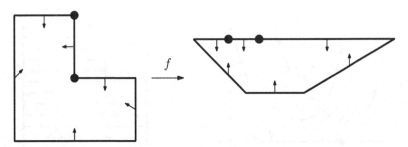

Figure 5.10. An oblique derivative problem on an L-shaped region and a map that suggests a solution.

onto a region bounded by appropriately oriented horizontal and vertical segments and then take the real part of this map.

In this section we consider a generalization of this idea: finding a harmonic function that satisfies *piecewise-constant homogeneous oblique derivative boundary conditions*. On side k of P we are given a direction $\theta_k \in (-\pi/2, \pi/2]$, measured clockwise from the interior normal of side k, in which the derivative of the solution ϕ must be zero. See Figure 5.10. This problem includes the case of section 5.3, for $\theta_k = 0$ is a homogeneous Neumann condition and $\theta_k = \pi/2$ is a constant Dirichlet condition. The more general oblique derivative problem also occurs in applications, as we demonstrate toward the end of this section.

One solution to the oblique derivative problem is $\phi = C$ for any constant C. Do other solutions exist? How can we find them?

As is suggested by Figure 5.10, the answers come from a simple extension of the ideas we have already applied. Suppose that we construct a conformal map $f(z)$ of P onto a domain Q such that along each side of Q, the oblique direction arrow for P is oriented in the vertical direction for Q (i.e., side k of Q makes an angle θ_k with the positive real axis). Then we set

$$\phi(z) = \operatorname{Re} f(z)$$

as usual. Along each boundary segment, the derivative of f in the indicated direction is now purely imaginary, and the derivative of ϕ is accordingly zero, as required. Thus ϕ solves the oblique derivative problem.

To construct f and Q, we proceed along now-familiar lines. First, the problem domain P is mapped to a canonical domain via a standard SC parameter problem; for our discussion, we continue to use the upper half-plane. From here a further SC integral with specified angles is then required to get to the target domain Q. As in the previous section, the side lengths of Q are not specified but are found as part of the solution.

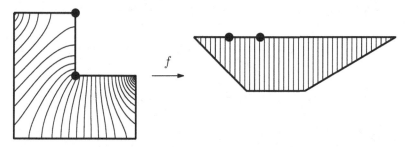

Figure 5.11. Solution of the L-shaped oblique derivative problem of Figure 5.10.

Figure 5.11 plots one result obtained in this way for the L-shaped example of Figure 5.10. The curves inside P are preimages of vertical lines in Q, which means that they are level curves of ϕ. Notice that here, as in Figure 5.10, the two vertices of P indicated by the heavy dots map to points along a side of Q that are not vertices. This occurs because the boundary condition does not change at these corners of P.

The discovered geometry of the image Q reveals information about the original problem. For example, in the present case, if $\phi = 0$ at the upper-left corner of P and $\phi = 1$ at the upper-right corner, we may compute from the position of the vertices along the boundary of Q that the values of ϕ at the six corners of P, beginning with the upper-left and proceeding counterclockwise, are 0, 0.2599, 0.5498, 1, 0.3267, and 0.1734. The reader who counts contours will note that eight or nine curves in the rightmost upper vertex of the L are so small as to be hidden under the thick line for the boundary; something like a fifth of the total "voltage drop" occurs in this millimeter or so of our figure.

Since all the boundary conditions of this problem are homogeneous, the solution is determined only up to a scale factor. We could multiply it by any constant (including -1, which would correspond to turning Q upside-down). Since $\phi = C$ is also a solution for any constant C, it is clear that there is a linear space of solutions of dimension at least 2. Could the dimension be larger? Such questions can be answered by thinking about other possible configurations for the target polygon Q. Evidently the angle at each vertex of Q is determined only up to multiples of π. This fact alone gives us a vector space of solutions to the oblique derivative problem whose dimension is countably infinite.

What shapes can we make by concatenating a horizontal segment, a diagonal at angle $\pi/6$ (mod π), another horizontal segment, and a diagonal at angle $3\pi/4$ (mod π)? As we saw in section 5.3 (see Figure 5.5), if vertices of Q are permitted to lie at infinity, there are numerous other possibilities besides

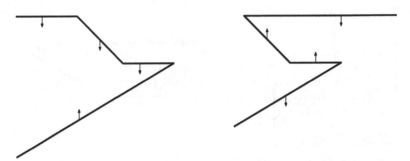

Figure 5.12. Alternative solutions to the L-shaped oblique derivative problem.

those of Figure 5.10, two of which are sketched in Figure 5.12. The prescription $\phi = \mathrm{Re}\, f$ applied to these new shapes will generate independent but unbounded solutions to the oblique derivative problem. Since we may permit $f(z) = \infty$ at any points along the boundary of P, the dimension becomes uncountable.

In applications, some of these unbounded solutions may be important, and one such example follows. But to simplify the discussion (and to keep in step with section 5.3), let us confine attention for the moment to bounded solutions ϕ that extend continuously to the boundary. The technique of solving such problems by SC mapping was investigated in a general way by Trefethen and Williams [TW86]. Here we suppose that the polygon P is first mapped to the upper half-plane so that the vertices map to $x_1 = \infty$ and $x_2 < x_3 < \cdots < x_n$ on the real axis. We also define $\theta_0 = \theta_n$ for convenience.

Theorem 5.3 ([TW86]). *A bounded and continuous solution ϕ to the oblique derivative problem attains its maximum and minimum values at at least one vertex x_k satisfying*

$$\theta_k < \theta_{k-1}, \qquad k = 1, \ldots, n \qquad (5.10)$$

(and possibly at other points as well). If κ is the number of such vertices, then the solutions form a vector space of dimension $\max\{1, \kappa\}$. If $\kappa \leq 1$, the only solutions are the constants. If $\kappa \geq 2$ then the solutions in the upper half-plane take the form

$$\phi(z) = C + \mathrm{Re}\left\{ e^{i\theta_n} \int_X^z p(\zeta) \prod_{k=2}^n (\zeta - x_k)^{-\pi^{-1}[(\theta_k - \theta_{k-1}) \bmod \pi]} d\zeta \right\}, \qquad (5.11)$$

where C is a real constant, p is an arbitrary real polynomial of degree at most $\kappa - 2$, X is any fixed point in (x_n, ∞), and the branch of each factor in the product is taken so that its value at X is positive.

Without attempting a formal proof, let us explain the reasoning behind this theorem. As we observed earlier, the target image Q must have sides making angles θ_k with the positive real axis, and we are free to add multiples of π to the interior angles of Q. This tells us that the interior angle between sides $k - 1$ and k must be

$$\theta_{k-1} - \theta_k + (q_k + 1)\pi, \qquad k = 2, \ldots, n,$$

for some integer q_k. Define

$$\beta_k = \frac{\theta_{k-1} - \theta_k}{\pi}.$$

The exponent in the SC integrand of (5.11) must therefore be $\beta_k + q_k + m_k$, where $m_k \geq 0$ is the multiplicity of x_k as a root of $p(\zeta)$. For the moment, let $k > 1$. Now $\beta_k \in (-1, 1)$, so to allow every case in which the map is continuous at x_k after integration, we choose $q_k = 0$ for $\beta_k \leq 0$ and $q_k = -1$ for $\beta_k > 0$. This selection is precisely what the (mod π) operation in (5.11) does.

However, we still must consider $x_1 = \infty$. To be consistent with $k > 1$, we define $\beta_1 = (\theta_n - \theta_1)/\pi$ and $q_1 = -\lceil \beta_1 \rceil$. Then the exponent of the SC singularity at x_1 must be $\beta_1 + q_1 + m_1$, for $m_1 \geq 0$. But the exponent implied by (5.11) is

$$-2 - \deg p - \sum_{k=2}^{n} (\beta_k + q_k) = \beta_1 - 2 - \deg p - \sum_{k=2}^{n} q_k.$$

Hence

$$-2 - \deg p - \sum_{k=1}^{n} q_k = m_1 \geq 0,$$

which, in light of (5.10), implies $\deg p \leq \kappa - 2$.

In geometric terms, each of the κ "naturally" obtuse angles ($\beta_k > 0$) of Q can be replaced by a smaller one. Since the "natural" turns add up to zero, two such reductions must be left alone just to make the turns sum to -2; the remaining $\kappa - 2$ reductions may be added back in anywhere in the form of slits, via p. (One may also choose to add back, say, just $\kappa - 4$ slits, which corresponds to complex roots of p and a multiple-sheeted Q.) The max–min property of the theorem is a straightforward geometric statement of how an extreme left or right point of Q must behave.

Note that Theorem 5.3 is similar to Theorem 5.2 of the last section; in fact, Theorem 5.3 is a true generalization. In Theorem 5.2 the number of Dirichlet values is equivalent to κ in the present context because the end of each Dirichlet interval corresponds to a drop from $\theta = \pi/2$ to $\theta = 0$. (However, since the Dirichlet values are not prescribed in the present context, no unique solution is singled out.) As before, the linear nature of the problem is captured by the polynomial p with arbitrary real coefficients, and the roots of p affect the geometry of the image Q. Depending on their locations, these roots may modify the angles of Q, introduce slits into it, or cause it to be multiple-sheeted. If we impose κ conditions on the solution (say, values at vertices satisfying (5.10)), then a linear system can be solved to determine p uniquely. Theorem 5.3 is also similar to Theorem 5.1 of section 5.2, but the insistence on *continuous* solutions here ruled out the possibility of different Dirichlet values on adjacent intervals.

Observe that each exponent in the integrand product inside (5.11) is in the range $(-1, 0]$. In the case $\kappa = 2$ (where p has no roots that might coincide with a vertex), this restriction implies that Q has at most one obtuse angle, at the image of x_1. All bounded solutions are combinations of constants and multiples of the real part of Q, so one simple figure describes the solution for $\kappa = 2$ completely. (Our L-shaped example of Figures 5.10 and 5.11 has $\kappa = 2$, for instance.) However, the geometric possibilities (as opposed to the dimension of the solution space) grow rapidly for $\kappa > 2$. Figure 5.13 illustrates some possibilities for a fixed $\kappa = 4$ problem on the L-shaped domain. Figure 5.14 presents solutions on other domains of problems with $\kappa = 2, 3, 4$.

One application of the oblique derivative problem comes from a problem in electronics, the classical (as opposed to quantum) **Hall effect**. Our discussion here is adapted from [TW86]. Under ordinary circumstances, the electrical current density **J** in a uniform plate flows in the direction of the electric field **E** (i.e., the negative of the gradient of the potential ϕ). Ignoring the constant of proportionality, we have $\mathbf{J} = \mathbf{E} = -\nabla\phi$, where ϕ satisfies the Laplace equation with constant Dirichlet and homogeneous Neumann boundary conditions on constant-voltage and insulated sides, respectively. If a uniform transverse magnetic field **B** is applied, however, Maxwell's equations imply that **J** and **E** are no longer collinear. In complex notation, we now obtain $E = (1 - iB)J$, and the two vectors meet at fixed angle $\alpha = \sin^{-1} |B|$ known as the **Hall angle**:

$$|J|/|E| = \cos\alpha, \quad \arg J - \arg E = \alpha.$$

This means that if we wish to determine the electric current distribution by solving a Laplace problem involving ϕ, the usual boundary conditions must be modified. On constant-voltage sides, ϕ will still be constant; we can enforce this condition as an oblique derivative boundary condition with $\theta_k = \pi/2$. On insulated sides, however, the condition that **J** must be orthogonal to the boundary

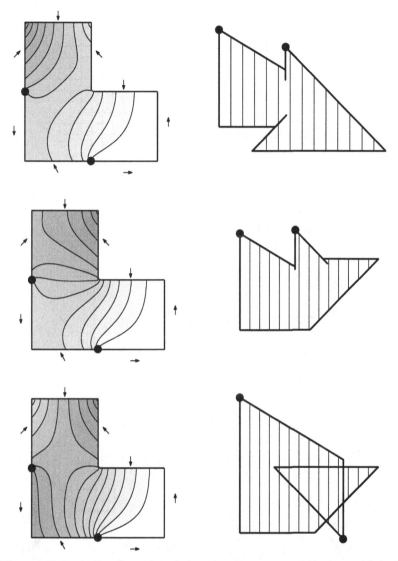

Figure 5.13. Examples of bounded solutions to a single oblique derivative problem of index $\kappa = 4$ on the L-shaped domain. The roots of the polynomial p in (5.11) determine the geometry of the image, whose real part gives the solution. Level curves are drawn, and on the left these are shaded from smallest (white) to largest (medium gray).

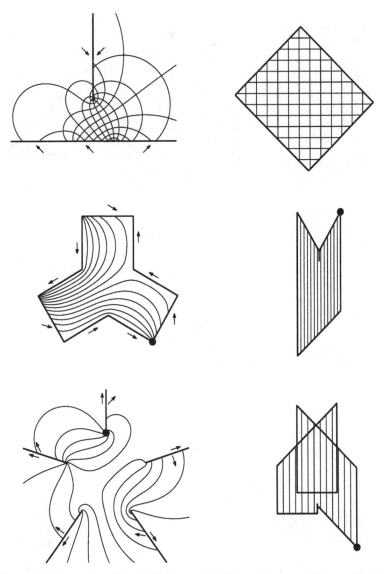

Figure 5.14. Examples of bounded solutions to oblique derivative problems of indices $\kappa = 2, 3, 4$ (top to bottom) on various domains. The arrows indicate the directions of zero derivative. Equipotentials (and, in the top case, flow lines) are shown.

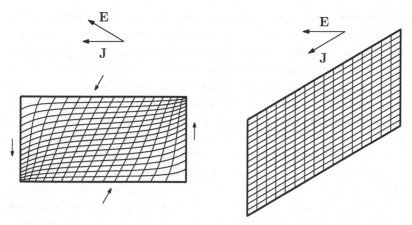

Figure 5.15. Illustration of the Hall effect by SC mapping. The presence of a transverse magnetic field in the rectangle on the left makes the electric field along insulated sides satisfy an oblique derivative condition with $\theta_k = \pi/6$, the Hall angle. In the parallelogram resulting from two SC maps (right), the equipotential and current flow lines are trivial, and these are readily transplanted back to the rectangle.

now implies that **E** does not point along the boundary but meets it at the angle α. Equivalently, ϕ satisfies the oblique derivative condition with angle $\theta_k = \alpha$.

The idea of solving such problems by SC maps goes back to Wick [Wic54] and was further developed by Versnel [Ver82, Ver83]. Figure 5.15 gives an example. As shown on the left, we consider a rectangular conductor of aspect ratio 2 with $\theta = \pi/2$ (i.e., applied potentials) specified on the left- and right-hand sides and $\theta = \pi/6$ on the top and bottom. For this problem, $\kappa = 2$, so by Theorem 5.3, there is one linearly independent nonconstant bounded solution. It follows that the parallelogram on the right of the figure represents the only target region we must consider. (The two free constants in the solution, corresponding to scaling and translating the parallelogram, allow us to match the values of the applied potentials.) The numerical SC map is straightforward, and the trivial equipotentials and current flow lines in the parallelogram are transplanted back to the rectangle, where they still meet everywhere at the angle $\pi/6$.

Our second example, **reflected Brownian motion** [HLS85, Wil95], illustrates an oblique derivative problem in which an unbounded solution is of interest. This is an idealized problem arising in queuing theory [Har78, New79] and storage theory [Wen82]. Our treatment is also taken from [TW86].

As in the last example, for simplicity we choose a rectangle of aspect ratio 2 for our geometry. Consider a particle moving at random in the rectangle. We may think of the position of the particle as representing the state of a pair of tandem queues: the x- and y-coordinates represent how full each of the queues is relative to its maximum allowable size. Customers arrive and are serviced at random at equal rates on average; this scenario is the origin of the Brownian motion. Suppose that if queue 1 becomes empty, it takes customers from queue 2. This arrangement corresponds to an oblique derivative boundary condition at angle $-\pi/4$ on the left-hand boundary. Similarly, suppose that if queue 2 becomes full, it transfers customers to queue 1. This stipulation corresponds to an oblique derviative boundary condition at angle $\pi/4$ on the top boundary. If queue 1 accepts no customers when full and queue 2 processes no customers when empty, we have Neumann boundary conditions on the other two boundaries. All together, we now have an oblique derivative boundary condition problem whose solution ϕ represents the probability density for the tandem system to lie in a particular state. See Figure 5.16. (The oblique directions do not coincide with the reflection directions for the wandering particle; those are obtained by reflecting the arrows through the normals.)

The index of this problem is $\kappa = 1$. Thus we see by Theorem 5.3 that the only bounded solution is a constant. There is one obvious additional condition that a physically appropriate solution to this problem must satisfy: since ϕ represents a probability density, its integral over the rectangle should be 1. This suggests

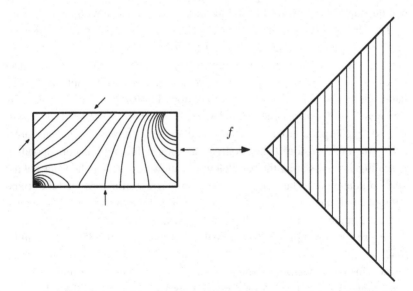

Figure 5.16. Illustration of reflected Brownian motion by SC mapping. A particle subject to Brownian motion in the rectangle at left, with oblique reflections at the boundaries (in directions parallel to reflections of the arrows through normals), leads to a steady-state probability distribution given by the real part of the map to the unbounded radial domain at right. Contours of constant probability (vertical lines at right) are transplanted back to the original domain.

that the solution should be the constant $\phi \equiv 0.5$. However, it can be shown that this solution has the property that a finite amount of probability "leaks out" of two vertices and reappears at the others—clearly inappropriate behavior.

An additional "corner condition" was derived by Newell [New79] to exclude this situation. In effect, this condition requires that the image polygon Q for our conformal map must be a radial domain (i.e., side k has constant argument θ_k). The constant solution violates the corner condition; therefore, we conclude that the physically meaningful solution is unbounded. We omit the details of the derivation, referring the reader to [TW86], but Figure 5.16 shows the result. SC mapping gives the solution to many digits of accuracy, and we may calculate, for example, that the mean position of the queue in steady state is $\langle x \rangle = 1.239964$, $\langle y \rangle = 0.482830$.

5.5 Generalized parameter problems

The main step of constructive Schwarz–Christoffel mapping is to find the unknown map parameters—namely, the prevertices. These parameters are uniquely determined by the geometry of the target region. In some applications, however, the geometry may be partially unknown, and instead one or more auxiliary conditions are to be imposed. Often this situation can be modeled by modifying the equations (such as (3.2)) that are used to determine the parameters, and the result can be called a **generalized parameter problem**. Such techniques arise often in design and inverse problems.

As an example, consider the problem of **resistor trimming** [Tre84]. In this problem, one has a polygonally shaped electrical resistor that is fabricated to a slightly lower resistance than is truly desired. A laser is to be used to cut a slit at a given location in order to adjust the resistance to the correct value. (See Figure 5.17.) How long should the slit be?

Recall (see section 2.5) that a given polygon, with four distinguished "corner" vertices, can be mapped to a rectangle of a specific aspect ratio, the conformal modulus. The modulus is equal to the resistance (in appropriate units) of the polygon. In the trimming problem, the geometry is partly unknown. This gives us the flexibility to impose the conformal modulus we want. Specifically, the vertex w_* at the tip of the slit is unknown. That removes two real equations from the side-length conditions (3.2). We replace them with the conditions:

- The sides incident on w_* must have equal length, and
- The conformal modulus must be a given number R.

In the numerical implementation described in section 4.3, the strip is used as an intermediate domain for the rectangle map, and the strip parameter L is easily

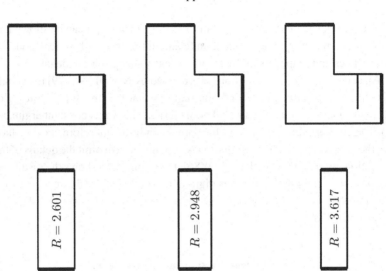

Figure 5.17. Resistor trimming problem. A slit of unknown length is to be cut in order to make the resistance (when voltage is applied on heavy lines) reach a target value. The SC parameter problem can be modified to impose the resistance (i.e., conformal modulus) condition directly and solve for the slit length.

linked to R. The resulting generalized parameter problem has a unique solution provided that the slit, if extended, separates the electrodes.

We can also see the resistor trimming problem as an inverse problem of determining the length of a straight crack in an insulator given a resistance measurement. This problem has applications in nondestructive testing, such as might arise for aircraft and nuclear reactors. A more realistic model might be to let the length, angle, and starting point of the crack all be unknown. For each geometric unknown, we require a resistance measurement using different corners (segments of applied voltage), as sketched in Figure 5.18. If one allows compound cracks of several segments, more measurements are needed. Not surprisingly,

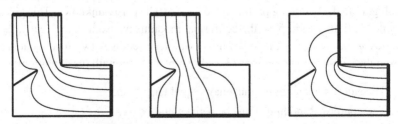

Figure 5.18. Multiple resistance measurements (applying voltage on heavy lines) can be used to determine multiple geometric unknowns, such as the starting point, angle, and length of a crack.

issues of existence and uniqueness become more difficult, as does convergence for the generalized parameter problem. See [EIN95, Nic97] for details.

A further variation on the idea is discussed by Elcrat and Hu [EH96]. Using doubly connected SC maps (as described in section 4.9), they show how to locate cracks that are interior to a polygon. Four resistance measurements are required to find the endpoints of a linear crack, and so on. At each iteration in solving the parameter problem, one must map the annulus to a rectangle with a current-aligned slit, for which the resistance is trivial to find.

5.6 Free-streamline flows

In this section we show that the SC idea can be applied to the classical fluid mechanics problem of ideal free-streamline flows. These flows arise in the study of wakes, jets and cavities. The use of complex analysis to study such flows dates back to 1868 and is due to Kirchhoff and Helmholtz [Hel68]. Good background references for these ideas are from [Mon83, ST84]; classic older references include [BZ57, Gil60, Gur65]. However, the classical conformal-mapping approach often leads to Riemann surfaces of unknown topology and is difficult to implement in generality. Elcrat and Trefethen [ET86] and Dias, Elcrat, and Trefethen [DET87] proposed a new, more direct approach that applies equally well to all polygonal regions, and these papers form the basis of the discussion here.

A typical situation is wake flow, depicted in Figure 5.19. An ideal incompressible fluid undergoes irrotational flow rightward past an obstacle Γ. The

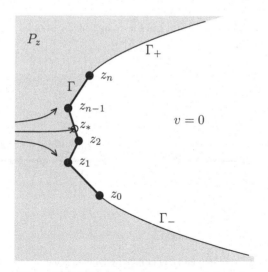

Figure 5.19. Wake flow past a polygonal obstacle.

obstacle consists of straight line segments with endpoints $z_0, \ldots, z_n$. Here $\alpha_k \pi$ is the angle made at z_k on the upstream side of Γ. At an unknown **stagnation point** z_* (assumed to be distinct from the vertices), the flow separates into upper and lower parts. As the fluid passes past the ends z_0 and z_n, it does not remain attached to the other side of the obstacle (that would require infinite acceleration); rather, it flows around a **wake** in which the (complex) fluid velocity $v(z)$ is zero. The curves Γ_+ and Γ_- separating the wake from moving fluid are unknown free streamlines determined by $|v(z)| = 1$, a condition deriving from the continuity of pressure.

Let P_z be the region of moving fluid. Because the flow is incompressible and irrotational in P_z, v can be interpreted as the gradient of a real, harmonic potential. In complex terms we can write

$$\bar{v}(z) = \frac{dw}{dz} \tag{5.12}$$

for a new complex variable $w(z)$. The quantity $\bar{v}$ is called the **hodograph** variable and is often given the symbol ζ. In the w-plane the obstacle is mapped to a slit, with the stagnation point mapping to the tip of the slit. See Figure 5.20. Without loss of generality, we assume that $w_* = w(z_*) = 0$. We make one further transformation that maps the slit plane P_w into the upper half-plane. This new variable x is given by

$$w = \tfrac{1}{2} W (x - x_*)^2, \tag{5.13}$$

where W and x_* are real constants chosen so that $x_0 = x(z_0) = -1$ and $x_n = x(z_n) = 1$. Recall that $x_* = x(z_*)$ is the image of the stagnation point, which is a priori unknown.

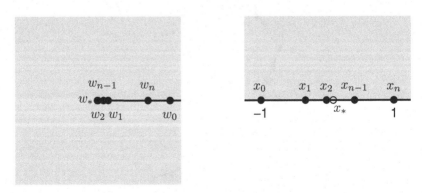

Figure 5.20. Velocity-potential and computational planes in the wake-flow problem.

We summarize what is known about the relationships between variables:

$$\arg(dz/dw) \text{ changes for } x \in (-1, 1) \quad \text{(polygonal obstacle)}, \quad (5.14\text{a})$$
$$\text{only at } x_1, \ldots, x_{n-1} \text{ and } x_*$$

$$|dz/dw| = 1 \text{ for } x \notin [-1, 1] \quad \text{(because of (5.12))}, \quad (5.14\text{b})$$

$$\arg(dz/dw) \text{ is prescribed at } x_n \quad \text{(obstacle orientation)}, \quad (5.14\text{c})$$

$$\arg(dz/dw) = 0 \text{ at } x = \infty \quad \text{(flow orientation)}. \quad (5.14\text{d})$$

The first of these conditions is the standard SC requirement. The second condition is new, however, and affects how we meet the first as well. Our goal is to express dw/dz as a product, each term of which introduces a jump in argument at a prevertex while affecting neither the other prevertices nor condition (5.14b). An elementary factor $g_k(x)$ of dw/dz should have the action depicted in Figure 5.21: the upper half-plane is mapped to the interior of a semicircle. Because x_k maps to 0, we get the requisite jump in argument at x_k (after exponentiation by $\alpha_k - 1$). Since the interval $[-1, 1]$ maps to the polygonal part of the boundary, the argument is otherwise constant there. Finally, because the rest of the real axis maps to the arc of the semicircle, $|g_k| = 1$ there.

To construct a formula for g_k, we first suppose temporarily that $x_k = 0$, and we set $y = g_k(x)$. The Joukowski transformation $(y + y^{-1})/2$ maps the interior of the unit upper semicircle in the y-plane to the lower half-plane, with the circular arc mapping to $[-1, 1]$ and the rest of the boundary mapping to the remainder of the real axis. Hence $x = 2/(y + y^{-1})$ describes g_k^{-1}. Solving for y, we find

$$y = \frac{x}{1 + \sqrt{1 - x^2}}.$$

If $x_k \neq 0$, we must first apply the Möbius transformation $(x - x_k)/(1 - x_k x)$

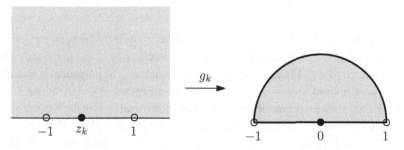

Figure 5.21. Elementary factor (5.15) in the SC map to a free-streamline flow region.

to map $\{-1, x_k, 1\}$ to $\{-1, 0, 1\}$. Putting all this together, we arrive at

$$g_k(x) = \frac{x - x_k}{1 - x_k x + \sqrt{(1 - x^2)(1 - x_k^2)}}, \tag{5.15}$$

and therefore

$$\frac{dz}{dw} = e^{i\gamma\pi} \left[\frac{x - x_*}{1 - x_* x + \sqrt{(1 - x^2)(1 - x_*)^2}} \right]^{-1}$$

$$\times \prod_{k=1}^{n-1} \left[\frac{x - x_k}{1 - x_k x + \sqrt{(1 - x^2)(1 - x_k)^2}} \right]^{\alpha_k - 1}. \tag{5.16}$$

The exponent of the x_* term comes from the fact that w_* is the tip of the slit in P_w, and the factor $e^{i\gamma\pi}$ ensures (5.14c) if $\gamma\pi$ is the angle of side n. To satisfy (5.14d), we allow $x \to \infty$ and derive a condition on x_*:

$$x_* = -\cos \left(\gamma\pi - \sum_{k=1}^{n-1} (\alpha_k - 1) \cos^{-1}(-x_k) \right). \tag{5.17}$$

Thus the stagnation point is determined by the other prevertices.

Because the right-hand side of (5.16) is expressed in x, we use (5.13) and apply the chain rule. This causes a cancellation of $(x - x_*)$, and we finally obtain

$$z(x) = A + W \int^x (x - x_*) \frac{dz}{dw} \, dx$$

$$= A + C \int^x \left(1 - x_* \xi + \sqrt{(1 - \xi^2)(1 - x_*)^2} \right)$$

$$\times \prod_{k=1}^{n-1} \left[\frac{\xi - x_k}{1 - x_k \xi + \sqrt{(1 - \xi^2)(1 - x_k)^2}} \right]^{\alpha_k - 1} d\xi. \tag{5.18}$$

Note that $C = W e^{i\gamma\pi}$ is a complex constant determined by the geometry of Γ.

The unknown SC parameters are the prevertices $x_1, \ldots, x_{n-1}$, and they are determined by $n - 1$ ratios of successive side lengths. (The polygonal boundary here is not closed, so $n - 3$ conditions do not suffice as they did in section 3.1.) Once they are found numerically, vertical and horizontal lines in the x-plane map via (5.18) to equipotentials and streamlines in the physical plane. Furthermore, the drag and lift forces on the object Γ are easily computed. By Bernoulli's formula, the pressure at a point $z \in \Gamma$ is equal to $\frac{1}{2}|v(z)|^2$ (assuming a suitable zero reference and unit fluid density). Therefore the force on a segment from z_a to z_b is

$$F_{a,b} = \frac{i}{2} \int_{z_a}^{z_b} v\bar{v}\, dz = \frac{i}{2} \int_{w_a}^{w_b} v(w)\, dw = \frac{i\, W}{2} \int_{x_a}^{x_b} (x - x_*)v(x)\, dx.$$

Since $|v| = 1$ on the free streamline Γ_-, it can be analytically continued into the lower-half x-plane via $\hat{v}(x) = 1/\bar{v}(\bar{x})$ (see [Gil60, pp. 350 and 370]). In other words,

$$F_{a,b} = \frac{i\, W}{2} \int_{x_a}^{x_b} \frac{(x - x_*)}{\bar{v}(x)}\, dx = \frac{i\, W}{2} \int_{x_a}^{x_b} (x - x_*)\frac{dz}{dw}\, dx.$$

The total force on Γ can be obtained by integrating over $[-1, 1]$. By Cauchy's theorem, we can integrate over any contour enclosing $[-1, 1]$:

$$F = F_{\text{drag}} + i\, F_{\text{lift}} = \frac{i\, W}{2} \oint (x - x_*)\frac{dz}{dw}\, dx.$$

The integrand here is identical to that in (5.18).

Observe that the integrand in (5.18) has a new type of singularity at ± 1 that is not of Gauss–Jacobi type. In fact, it can be shown that if $H(z)$ is the integrand, then $H(z) = h(\sqrt{z \pm 1})$ for an analytic function h. By a change of variable, though, we find that

$$\int_{\pm 1}^{z} H(\zeta)\, d\zeta = \int_{0}^{\sqrt{z \pm 1}} 2\eta h(\eta)\, d\eta,$$

which is now treatable by Gauss–Legendre quadrature.

All the computational essentials were implemented by Elcrat and Trefethen in the FORTRAN package KIRCH1 [ET86], available from Netlib. Examples of wake flows are shown in Figure 5.22.

Jet flows can be handled similarly. In this case, the flow is between two semi-infinite polygonal walls, and the free streamlines bound the exiting jet of fluid. The image of the flow region in the w-plane has two parallel slits instead of just one slit, but this fact changes the details only slightly [DET87].

5.7 Mesh generation

In this small section we comment on the advantages, disadvantages, and possibilities of SC maps in mesh generation. However, neither author is an expert in

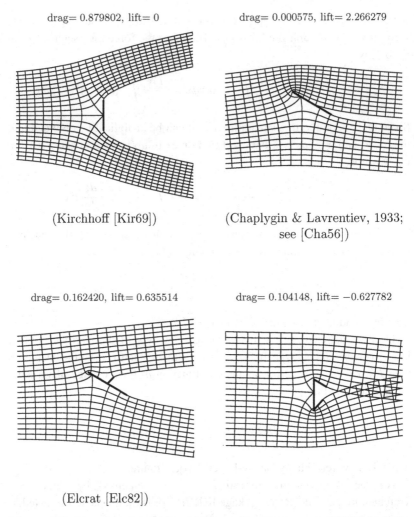

drag= 0.879802, lift= 0   drag= 0.000575, lift= 2.266279

(Kirchhoff [Kir69])   (Chaplygin & Lavrentiev, 1933;
                                                see [Cha56])

drag= 0.162420, lift= 0.635514   drag= 0.104148, lift= −0.627782

(Elcrat [Elc82])

Figure 5.22. Examples of wake flows (data produced by program KIRCH1). At top left, the drag is known to be exactly $2\pi(4 + \pi) = 0.87980169\ldots$. At top right, the separation point is specified on the middle of the plate; at bottom left, it is at the tip of the spoiler. At bottom right, the flow nonphysically crosses itself.

this large and well-developed field, so we urge the interested reader to consult a text on mesh generation for a more complete discussion [Lis99, TSW99].

Let x and y be real variables in the physical domain, and let ξ and η be orthogonal real variables in a computational domain with trivial geometry (usually a rectangle or disk). The goal of structured, boundary-conforming mesh generation is to describe a coordinate transformation $(x(\xi, \eta), y(\xi, \eta))$ that transplants a differential equation from the physical domain to a transformed version in ξ and η. The transformation may be written in differential form as

$$\begin{bmatrix} dx \\ dy \end{bmatrix} = J(\xi, \eta) \begin{bmatrix} d\xi \\ d\eta \end{bmatrix},$$

where $J(\xi, \eta)$ is the Jacobian matrix, which in general depends on location. An orthogonal transformation is one for which

$$J = \begin{bmatrix} h(\xi, \eta) & 0 \\ 0 & g(\xi, \eta) \end{bmatrix} \begin{bmatrix} \cos\theta(\xi, \eta) & -\sin\theta(\xi, \eta) \\ \sin\theta(\xi, \eta) & \cos\theta(\xi, \eta) \end{bmatrix}.$$

From a partial differential equations viewpoint, the chief advantage of an orthogonal transformation is that no "cross terms" are introduced in the Laplacian operator. (Although orthogonality is not as important as it once was, at least near-orthogonality may still be useful.)

A conformal transformation is an orthogonal transformation that additionally satisfies $g = h$; that is,

$$J = h(\xi, \eta) \begin{bmatrix} \cos\theta(\xi, \eta) & -\sin\theta(\xi, \eta) \\ \sin\theta(\xi, \eta) & \cos\theta(\xi, \eta) \end{bmatrix}. \tag{5.19}$$

In other words, the "metric" or stretching of the transformation does not depend on the direction of $d\zeta = d\xi + i d\eta$.[3] In complex terms, if we define $z = x + iy$ and $\zeta = \xi + i\eta$, then (5.19) is just a statement of the Cauchy–Riemann equations.[4]

Conformal structure makes transplantation especially simple to describe. For example, if we write $z = f(\zeta)$, then the Laplacian operator transforms according to

$$\Delta_z = |f'(\zeta)|^{-2} \Delta_\zeta. \tag{5.20}$$

Moreover, Dirichlet and homogeneous Neumann boundary conditions are preserved under such a transplantation. In many contexts this fact implies that fast Poisson solvers may be used to great advantage after transplantation to ζ.

There are disadvantages to conformal methods, however. An inspection of many of the figures in this book makes it clear that the size of equally spaced

---

[3] According to Ives [Ive82], one of the motivations for using conformal mapping in the early years of computational mesh generation was the economy of storing just one metric quantity at each grid point.

[4] For a delightfully clear and detailed examination of these connections, in which (5.19) is described as an "amplitwist," see Chapter 4 of [Nee97].

cells in the computational domain can vary greatly and rapidly in the physical domain (see in particular Figure 2.8). This variation makes a certain amount of physical sense, at least in some contexts, but is undesirable in many applications. A particular problem with (5.20) in SC mapping is the presence of singularities in f' due to the corners. As a mapping, the SC formula handles corners elegantly and completely, but here they remain a significant challenge in the computational domain. Finally, of course, conformal mapping is an inherently two-dimensional method, and much of the need for mesh generation is in three dimensions.

Ives suggests that conformal maps are useful as *part* of the mesh generation process [Ive82]. For example, one can use a conformal map to find a boundary correspondence between computational and physical regions, followed by an algebraic or orthogonal (but not conformal) extension to the interior. This type of mapping could allow more control over cell sizes than strict conformal mapping. Ives also reported success in "stacking" orthogonal grids and connecting them orthogonally through a third dimension. We suspect, however, that the role of SC mapping in mesh generation is not likely to expand in the future.

5.8 Polynomial approximation and matrix iterations

Approximation in the complex plane is a highly developed subject (see [Wal64] and [Gai87] for classic and modern references). In this section we point out how SC maps can be used to actually construct one important tool, the Faber polynomials, and we illustrate the connections between these and the major computational field of iterative methods for solving linear systems.

Let D be a connected set in the plane with complement D^c that is simply connected in the extended plane (on the Riemann sphere). By the Riemann mapping theorem, there exists a conformal map $\Phi(z)$ from D^c to the exterior of the unit circle such that $\Phi(\infty) = \infty$. (See Figure 5.23.) Now Φ has the Laurent expansion

$$\Phi(z) = C^{-1}(z + a_0 + a_1 z^{-1} + a_2 z^{-2} + \cdots), \qquad (5.21)$$

where $|C| > 0$ is the capacity of D.[5] The **level curves** of D are the images under Φ^{-1} of the circles of radius $R > 1$.

When D is the interior of a bounded polygon, the SC exterior map described in section 4.4 gives us access to Φ^{-1}. If we let $w = \Phi(z)$, $u = 1/w$, and let f

---

[5] Normally one determines Φ uniquely by requiring $C > 0$, but relaxing this condition allows us to more easily match the discussion in section 4.4.

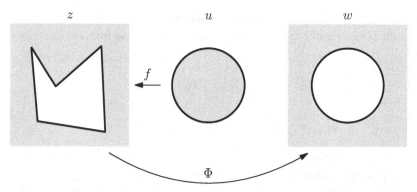

Figure 5.23. Notation for the approximation discussion for simply connected regions. The map f is a Schwarz–Christoffel map.

be the SC disk-to-exterior map (see (4.6))

$$f(u) = A - C \int^u \omega^{-2} \prod_{k=1}^n \left(1 - \frac{\omega}{u_k}\right)^{1-\alpha_k} d\omega, \qquad (5.22)$$

then $\Phi^{-1}(w) = f(1/w)$ and $|C|$ is indeed the capacity, as was pointed out in section 4.4.

Referring again to (5.21), we see that the expansion of $[\Phi(z)]^m$ has a leading (analytic) polynomial part of degree m. This polynomial $\phi_m(z)$ is called the **Faber polynomial** of degree m for the region D. With an SC map we can find the coefficients of the Faber polynomials [SV93]. We write

$$z = \Phi^{-1}(w) = Cw + b_0 + b_1 w^{-1} + b_2 w^{-2} + \cdots, \qquad (5.23)$$

$$\frac{dz}{dw} = C - b_1 w^{-2} - 2b_2 w^{-3} - \cdots. \qquad (5.24)$$

From (5.22) we also have

$$\frac{dz}{dw} = -w^{-2} f'(w^{-1}) = C \prod_{k=1}^n (1 - (wu_k)^{-1})^{1-\alpha_k}$$

$$= C \prod_{k=1}^n \left(1 + \frac{\alpha_k - 1}{u_k} w^{-1} + \gamma_{k,2} w^{-2} + \cdots\right),$$

where $\gamma_{k,2}$ and higher terms can be found from the binomial theorem. We can match this expression against (5.24); the residue condition (4.7) ensures that the w^{-1} term in dz/dw vanishes. This comparison determines all the coefficients in the expansion (5.23) except the constant term b_0, which can be estimated

accurately by evaluating f near the origin and subtracting off the known part
of the series. Finally, the Faber polynomials can be found from the recursion

$$\phi_0 = 1,$$

$$\phi_1 = \frac{1}{C}(z - b_0),$$

$$\phi_{m+1} = \frac{1}{C}\left(z\phi_m - (b_0\phi_m + \cdots + b_m\phi_0) - mb_m\right).$$

The level curve $|\Phi(z)| = 1$ is, of course, the boundary of D. Faber poly-
nomials approximate powers of Φ; therefore, the lemniscates $|\phi_m(z)| = 1$ ap-
proximate the boundary of D. In fact, these curves tend to oscillate around the
boundary, as shown in Figure 5.24. Because the Faber polynomials have nearly
constant modulus on the boundary, by Rouché's Theorem [Ahl78, Nee97] they
are "nearly minimax" on D. For example, when D is the complement of a finite
interval, the Faber polynomials are just twice the Chebyshev polynomials. For
more on these topics, see [Cur71, Ell83], for example.

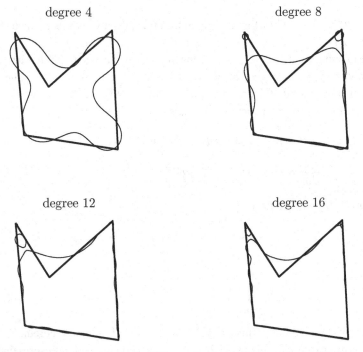

degree 4   degree 8

degree 12   degree 16

Figure 5.24. Level curves $|\phi_m| = 1$ of Faber polynomials. Because $\phi_m \approx \Phi^m$, the curves
approximate the polygon. The re-entrant corner (with respect to the interior) is the last
to be resolved.

These concepts have an important application in the area of Krylov subspace methods for solving linear systems. To solve $Ax = b$, a Krylov subspace method finds a sequence of approximations x_m such that

$$b - Ax_m = p_m(A)b,$$

where p_m is a polynomial of degree m characterizing the method and satisfying $p_m(0) = 1$. The object is to make the residual norm $\|b - Ax_m\|$ small. This quantity is linked to the size of $p_m(z)$ on the spectrum of A. Hence Krylov methods are connected to polynomial minimax problems. If the spectrum is well approximated by D, then the preceding map Φ above leads to an estimate of the convergence rate of the Krylov method. Specifically, if $0 \notin D$ (the system is nonsingular), then

$$\lim_{m \to \infty} \|b - Ax_m\|^{1/m} \approx \frac{1}{|\Phi(0)|} . \tag{5.25}$$

Furthermore, the Faber polynomials can be used to construct the polynomials p_m of the Krylov method, given an estimate of the spectrum [SV93]. A similar approach using SCPACK was presented in [Li92]. For more about the connections between matrix iterations and complex approximation, see [DTT98, Gre96, Nev93]. In particular, [DTT98] discusses six factors that affect the quality of approximation in (5.25). Examples of the use of SC maps to construct and analyze matrix iterations include [DTT98, HKH94, Kos94, Li92, Sta93, SV93].

If D^c is not simply connected, there is no analytic map Φ to the exterior of a disk. However, we can make some progress in the case where D is symmetric about the real axis, as discussed in the next section.

5.9 Symmetric multiply connected domains

In this last section we consider the problem of finding the Green's function in a symmetric, multiply connected region—which, we shall see, is also a problem of conformal mapping. As in section 5.2, we define the Green's function for D as the function $g(z)$ satisfying

$$\Delta g(z) = 0 \qquad \text{for } z \in D^c,$$

$$g(z) \to 0 \qquad \text{for } z \to D,$$

$$g(z) \sim \log |z| \qquad \text{for } z \to \infty.$$

The case in which D consists of two disjoint intervals on the real axis was investigated in the 1930s by Akhiezer, who used elliptic functions [Akh56]. For more than two such intervals, Widom [Wid69] derived results in 1969 using SC maps. The case of intervals has applications to digital filtering [PM72, SSW01].

Following the method outlined by Widom, Green's functions for domains D consisting of κ disjoint polygons symmetric about the real axis were constructed numerically by Embree and Trefethen [ET99]. The first step of their procedure was to use symmetry to reduce the problem to that part of the exterior of D lying in the upper half-plane. The symmetry implies the Neumann condition $dg/dn = 0$ for those portions of the real axis between components of D.

This step reduces the problem to the type discussed in section 5.3. However, Theorem 5.2 does not apply—in fact, we find there that the integrand polynomial p is zero and that the solution is constant. This is indeed the only bounded solution. The Green's function, on the other hand, has log $|z|$ behavior at infinity, which requires the target geometry to extend infinitely to the right. As illustrated in Figure 5.25, the correct target image is a semi-infinite strip with $\kappa - 1$ slits, one for each of the real intervals between components of D. The locations of these slits must be chosen so that the vertical (Dirichlet) sides all line up. This configuration is still a linear problem of dimension $\kappa - 1$; full details of the process and its applications appear in [ET99].

Two more examples, of connectivity three and five, are shown in Figure 5.26. The second of these examples seems quite remarkable, in that one level curve self-intersects at four points. This feat was accomplished by adjusting the spaces between the squares, which have widths 1, 2, 3, 4, and 5. Since the overall scaling is arbitrary, three gaps need to be determined by solving three nonlinear equations derived from the geometric condition that all slits in the analog of the middle picture of Figure 5.25 have the same length. This description is an example of a generalized parameter problem (section 5.5).

In section 5.8 we used the map Φ from the exterior of a simply connected D to the exterior of the disk. For a symmetric multiply connected D, we may define $\Phi(z) = e^{G(z)}$, where G is the map from D^c to the slit strip (i.e., Re G is the Green's function). Now Φ maps half of D^c to half of a "spiked disk," as illustrated in Figure 5.25. The spikes are images of the segments of the real axis between components of D. Upon reflection across the real axis, we can continue Φ to the exterior of a full spiked disk. However, we can continue the map just as well by reflecting across the spikes. This implies that Φ must be multivalued in the full domain (i.e., it is a conformal map of a Riemann surface lying about D^c to the exterior of the spiked disk). However, this multivaluedness does not affect $|\Phi|$ (after all, the Green's function Re G is unique), and we can meaningfully draw level curves in the disk domain.

Furthermore, we still find that $|z/\Phi(z)|$ converges to the capacity of D as $z \to \infty$, and the asymptotic Krylov convergence factor in (5.25) is still appropriate.

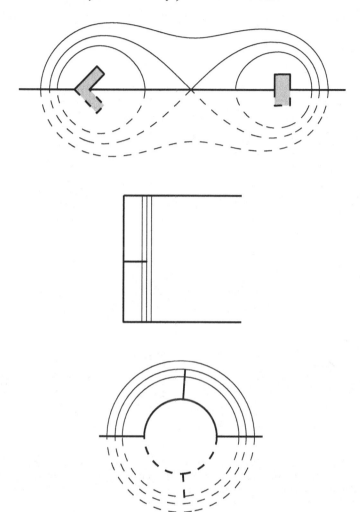

Figure 5.25. Finding the Green's function for a symmetric multiply connected region. Using symmetry, half of the region (top, exterior of shaded polygons) is mapped, via a standard plus a linear parameter problem similar to that in Theorem 5.2, to a semi-infinite slit strip (middle). The Green's function is the real part of this map. The strip can be exponentiated to give half of a spiked disk (bottom), which can be continued (in a multivalued way) by reflection. Three level curves of the Green's function are drawn in each domain.

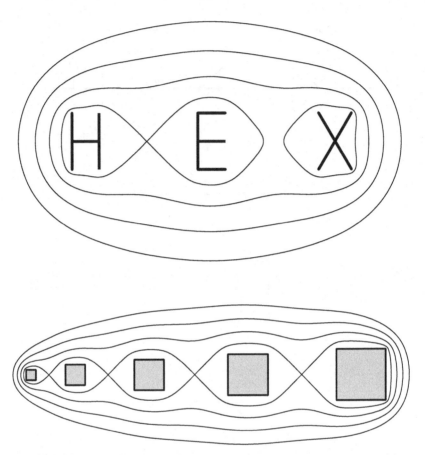

Figure 5.26. Examples of Green's functions for domains of connectivity three and five. Level curves are shown for each region. In the bottom picture, the gaps between squares were determined by solving a generalized parameter problem in order to make the four self-intersections occur.

One can also find the **harmonic measure** of each component of D with respect to the point at infinity. This quantity, which can be interpreted as the proportion of a minimal-energy charge distribution over D on each component, is simply given by the relative angles between spikes (and the real axis) in Figure 5.25. Moreover, it is possible using Φ to generate good polynomial approximations to any analytic function defined on each component of D—not just the zero function as in the minimax problem. In digital filters, one can use this idea to define one component as a "pass band" and another as a "stop band," for example. For more see [ET99, SS99, SSW01].

APPENDIX

Using the SC Toolbox

Most of the figures in this book were produced using the SC Toolbox for MATLAB.[1] This software is in the public domain and is available (at the time of this writing) from

$$\texttt{http://www.math.udel.edu/\~driscoll/SC/}$$

The toolbox is capable of half-plane and disk maps, exterior maps (section 4.4), strip maps (section 4.2), rectangle maps (section 4.3), and disk maps using the cross-ratio formulation (section 3.4). By tinkering with the provided routines, it is not too hard to produce maps to gearlike regions (section 4.8) and Riemann surfaces (section 4.7). Other variations and applications require more extensive programming efforts.

The toolbox defines polygons and the maps to them as named objects. Once created, these objects can be manipulated by using of common MATLAB functions and notations that have been extended to understand them. The main examples are

`plot(p)`, `plot(f)`	Plot polygon p or map f.
`eval(f,zp)` or `f(zp)`	Evaluate map f at point(s) zp.
`evalinv(f,wp)` or `fi=inv(f);fi(wp)`	Evaluate the inverse of f at point(s) wp.

The user may also examine and extract the data defining these objects and call low-level routines.

---

[1] Version 2.1 of the SC Toolbox and versions 5.2–6.0 of MATLAB. The first author hopes to maintain compatibility with future versions of MATLAB indefinitely. MATLAB is a registered trademark of The Mathworks, Inc., which has no affiliation with or responsibility for the SC Toolbox.

115

A polygon is created either from a list of its vertices (and its angle parameters, if it is unbounded) or from an interactive drawing using the function `polyedit`. The drawing program understands infinite vertices and allows points to be snapped to a grid or angles and side lengths to be quantized.

Given a polygon, one can construct a map to a region defined by it. The functions that accomplish this are named partly based on the canonical domain and map type and always ending with the letters `map`. These functions set up numerical quadrature data and solve the appropriate parameter problem using side lengths (section 3.1) or cross-ratios (section 3.4).

All the major toolbox functions can be accessed either by the command line, for maximum control and repeatability, or by a graphical user interface (GUI) activated by typing `scgui` at the prompt. The GUI gives the user access to virtually all of the toolbox's capabilities without any knowledge of MATLAB programming or syntax.

A user's guide is distributed with the toolbox and always has the most up-to-date information on its usage. In the following pages we show how some of the figures in this book were produced. (We omit purely cosmetic changes such as setting the thickness of curves.)

```
p = polygon([1+i,-1+i,-1-i,1-i]);
f = diskmap(p);
f = center(f,0);
plot(exp(i*linspace(0,2*pi,180)));
hold on, axis equal
[X,Y] = meshgrid((-4:4)/5,...
    (-100:100)/100);
plot(evalinv(f,X+i*Y),'k')
[X,Y] = meshgrid((-100:100)/100,...
    (-4:4)/5);
plot(evalinv(f,X'+i*Y'),'k')
```

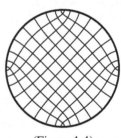

(Figure 1.4)

```
p = polygon([i,-i,Inf],[3/2,1/2,-1]);
f = hplmap(p);
axis([-3 3 -1.5 4.5]), hold on
plot(f,0.7*(-10:6),0.7*(1:12))
```

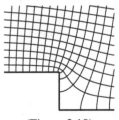

(Figure 2.10)

```
p = polygon([-4-i,4-i,4+i,-4+i]);
f = diskmap(p);
f = center(f,0);
plot(f,0.2*(1:4),angle(prevertex(f)))
```

(Figure 2.13)

```
w = [Inf,-1-i,-2.5-i,Inf,2.4-3.3i,...
     Inf,2.4-1.3i,Inf,-1+i,-2.5+i];
alf = [0,2,1,-.85,2,0,2,-1.15,2,1];
p = polygon(w,alf);
f1 = stripmap(p,[6 8]);
f2 = stripmap(p,[4 8]);
axis([-4.3 5.7 -6.15 3.85]), hold on
plot(f1,0,8)
pause, cla
plot(f2,0,8)
```

(Figure 4.6)

```
p = polygon([-5-i,-5-3i,5-3i,...
     5+i,5+3i,-5+3i]);
f = rectmap(p,[1 2 4 5]);
plot(f)
```

(Figure 4.8)

```
p = i*polygon([-0.5,1-1.5i,-0.5,0.5+2i]);
f = extermap(p);
axis([-3.05 2.8 -2.3 2.5]), hold on
plot(f,(4:9)/10,0)
```

(Figure 4.10)

```
f = rectmap(drawpoly);
r = rectangle(f);
M = pi/max(imag(r));
a = exp(min(real(r))*M);
b = exp(max(real(r))*M);
rad = log(linspace(a,b,8))/M;
plot(f,rad(2:end-1),6);
H = copyobj(get(gca,'child'),gca);
for n=1:length(H),
  set(H(n),'xdata',-get(H(n),'xdata'))
end
set(H,'linesty','--'), axis auto
```

(Figure 4.14)

```
p = polygon([1+i,1+2i,Inf,-.705+.971i,...
    Inf,-1-i,Inf,.705-.971i,Inf],...
    [2,1,-.3,2,-.7,2,-.3,2,-.7]);
f = center(diskmap(p),0);
axis(5.5*[-1 1 -1 1]), hold on
h = plot(f,0.7:0.05:0.95,0);
for n=1:length(h)
  w = (get(h(n),'xd') + i*get(h(n),'yd'));
  set(h(n),'xd',real(1./w),'yd',imag(1./w))
end
h = findobj(gca,'color',[0 0 1]);
for n=1:length(h)
  w = get(h(n),'xd') + i*get(h(n),'yd');
  if abs(w(1)) > abs(w(2)), w=w([2 1]); end
  u1 = w(1) + linspace(0,2*diff(w),100);
  u2 = w(1) + linspace(2*diff(w),100*diff(w),100);
  plot(1./[u1 u2]);
end
delete(h)
axis auto
```

(Figure 4.16)

```
p = polygon([i 0 2 2+i]);
theta = [1/2 1/6 1/2 1/6];
f1 = hplmap(p);
thetas = theta([end 1:end-1]);
alf = -mod(theta-thetas,1)+1;
f2 = hplmap(prevertex(f1),alf,i);
Q = polygon(f2); w = vertex(Q);
m = min(imag(w)); M = max(imag(w));
a = max(real(w)); t = [.01 1:99 99.99]/100;
plot(p), hold on
[X,Y] = meshgrid((1:15)*a/16,t*m);
plot(f1(evalinv(f2,X+i*(Y+M*X/a))),'k')
[X,Y] = meshgrid(t*a,(1:15)*m/16);
plot(f1(evalinv(f2,X'+i*(Y'+M*X'/a))),'k')
```

(Figure 5.15)

```
f = extermap(drawpoly);
phi = faber(f,8);
[x,y] = meshgrid(linspace(-4,4,80));
w = polyval(phi{9},x+i*y);
contour(x,y,abs(w),[1 1]);
hold on
plot(polygon(f))
```

(Figure 5.24)

Bibliography

[ADHE82] O. L. Anderson, R. T. Davis, G. B. Hankins, and D. E. Edwards. Solution of viscous internal flows on curvilinear grids generated by the Schwarz–Christoffel transformation. *Appl. Math. Comput.*, 10:507–24, 1982.

[AF97] M. J. Ablowitz and A. S. Fokas. *Complex Variables: Introduction and Applications*. Cambridge University Press, New York, 1997.

[Ahl78] L. V. Ahlfors. *Complex Analysis*. McGraw–Hill, New York, 3rd edition, 1978.

[Akh56] N. Akhiezer. *Theory of Approximation*. Ungar, New York, 1956.

[And75] R. Anderson. Analogue-numerical approach to conformal mapping. *Proc. IEE*, 122:874–6, 1975.

[Ban00] L. Banjai. Recursive algorithms in complex analysis. Unpublished honors project in Mathematics and Computation, Oxford University, 2000.

[BE92] M. Bern and D. Eppstein. Mesh generation and optimal triangulation. In D. Z. Du and F. Hwang, editors, *Computing in Euclidean Geometry*. World Scientific, Singapore, 1992. Also Xerox Palo Alto Research Center Technical report CSL-92-1.

[BEMR94] H. R. Baum, O. A. Ezekoye, K. B. McGrattan, and R. G. Rehm. Mathematical modeling and computer simulation of fire phenomena. *Theoret. Comput. Flu. Dyn.*, 6:125–39, 1994.

[BF81] P. L. Butzer and F. Fehér, editors. *E. B. Christoffel: The Influence of His Work on Mathematics and the Physical Sciences*. Birkhäuser, Boston, 1981.

[BG87] P. Bjørstad and E. Grosse. Conformal mapping of circular-arc polygons. *SIAM J. Sci. Stat. Comput.*, 8:19–32, 1987.

[Bin61] K. J. Binns. The magnetic field and centring force of displaced ventilating ducts in machine cores. *Proc. IEE*, 108 C:64–70, 1961.

[Bin62] K. J. Binns. Pole-entry flux pulsations. *Proc. IEE*, 109 C:103–7, 1962.

[Bin64] K. J. Binns. Calculations of some basic flux quantities in induction and other doubly-slotted electrical machines. *Proc. IEE*, 111:1847–58, 1964.

[BL63] K. J. Binns and P. J. Lawrenson. *Analysis and Computation of Electric and Magnetic Field Problems*. Pergamon, New York, 1963.

121

[BP93]   M. Brady and C. Pozrikidis. Diffusive transport across irregular and fractal walls. *Proc. Roy. Soc. Lond. A*, 442:571–83, 1993.

[BRK79]   K. J. Binns, G. R. Rees, and P. Kahan. Evaluation of improper integrals encountered in the use of conformal transformation. *Inter. J. Numer. Meth. Eng.*, 14:567–80, 1979.

[Bro81]   P. R. Brown. A non-interactive method for the automatic generation of finite element meshes using the Schwarz–Christoffel transformation. *Comput. Meth. Appl. Mech. Eng.*, 25:101–26, 1981.

[BZ57]   G. Birkhoff and E. H. Zarantonello. *Jets, Wakes, and Cavities*. Academic Press, New York, 1957.

[CCBS00]   S. Costa, E. Costamagna, P. Di Barba, and A. Savini. An innovative application of numerical Schwarz–Christoffel transformations to the optimal shape design of a permanent magnet motor. In *6th International Workshop on Optimization and Inverse Problems in Electromagnetism*, Torino, Italy, 2000. Digests.

[CF98]   E. Costamagna and A. Fanni. Computing capacitances via the Schwarz–Christoffel transformation in structures with rotational symmetry. *IEEE Trans. Magnetics*, 34:2497–500, 1998.

[Cha56]   S. A. Chaplygin. *Selected Works On Wing Theory*. Garbell Research Foundation, San Francisco, 1956. English translation from the original Russian by M. A. Garbell.

[Cha89]   M. A. Chaudhry. *An Extended Schwarz–Christoffel Transformation for Analysis and Design of High Capacity Integrated Circuit Lines and Resistors*. PhD thesis, Department of Electrical and Computer Engineering, University of California, Irvine, 1989.

[Cha92]   M. A. Chaudhry. An extended Schwarz–Christoffel transformation for numerical mapping of polygons with curved segments. *COMPEL*, 11:277–93, 1992.

[Chr67]   E. B. Christoffel. Sul problema delle temperature stazonarie e la rappresentazione di una data superficie. *Ann. Mat. Pura Appl. Serie II*, 1:89–103, 1867.

[Chr70a]   E. B. Christoffel. Sopra un problema proposto da Dirichlet. *Ann. Mat. Pura Appl. Serie II*, 4:1–9, 1870.

[Chr70b]   E. B. Christoffel. Über die Abbildung einer *n*-blattrigen einfach zusammenhängender ebenen Fläche auf einen Kreise. *Göttingen Nachrichten*, pp. 359–69, 1870.

[Chr71]   E. B. Christoffel. Über die Integration von zwei partiellen Differentialgleichungen. *Nachr. Kgl. Ges. Wiss. Göttingen*, pp. 435–53, 1871.

[Cos87]   E. Costamagna. On the numerical inversion of the Schwarz–Christoffel conformal transformation. *IEEE Trans. Microw. Theory Tech.*, 35:35–40, 1987.

[Cos01]   E. Costamagna. Numerical inversion of the Schwarz–Christoffel conformal transformation: Strip-line case studies. *Microwave Opt. Tech. Lett.*, 28:179–83, 2001.

[CS92]   M. A. Chaudhry and R. Schinzinger. Numerical computation of the Schwarz–Christoffel transformation parameters for conformal mapping of arbitrarily shaped polygons with finite vertices. *COMPEL*, 11:263–75, 1992.

[CST95]   J. C. Cockburn, Y. Sidar, and A. R. Tannenbaum. Stability margin optimization via interpolation and conformal mappings. *IEEE Trans. Aut. Cont.*, 40:1066–70, 1995.

[Cur71]   J. H. Curtiss. Faber polynomials and the Faber series. *Amer. Math. Monthly*, 78:577–96, 1971.

[CZ75]   L. A. Cherednichenko and I. K. Zhelankina. Determination of the constants that occur in the Christoffel–Schwarz integral. *Izv. Vyssh. Ucebn. Zaved. Elektromehanika*, 10:1037–40, 1975.

[Däp87]   H. Däppen. Wind-tunnel wall corrections on a two-dimensional plate by conformal mapping. *AIAA J.*, 25:1527–30, 1987.

[Däp88]   H. Däppen. *Die Schwarz–Christoffel-Abbildung für zweifach zusammenhängende Gebiete mit Anwendungen*. PhD thesis, ETH Zürich, 1988.

[Dav79]   R. T. Davis. Numerical methods for coordinate generation based on Schwarz–Christoffel transformations. In *4th AIAA Comput. Fluid Dynamics Conf.*, pp. 1–15, Williamsburg, VA, 1979.

[DE92]   F. Dias and A. R. Elcrat. Ideal jet flow with a stagnation streamline. *European J. Mech. B*, 11:233–47, 1992.

[DE93]   T. K. DeLillo and A. R. Elcrat. Numerical conformal mapping methods for exterior regions with corners. *J. Comput. Phys.*, 108:199–208, 1993.

[DeL87]   T. K. DeLillo. On some relations among numerical conformal mapping methods. *J. Comp. Appl. Math*, 19:363–77, 1987.

[DEP01]   T. K. DeLillo, A. R. Elcrat, and J. A. Pfaltzgraff. Schwarz–Christoffel mapping of the annulus. *SIAM Rev.*, 43:469–77, 2001.

[DET87]   F. Dias, A. R. Elcrat, and L. N. Trefethen. Ideal jet flow in two dimensions. *J. Fluid Mech.*, 185:275–88, 1987.

[Dri96]   T. A. Driscoll. A MATLAB toolbox for Schwarz–Christoffel mapping. *ACM Trans. Math. Soft.*, 22:168–86, 1996.

[DS96]   J. E. Dennis, Jr., and R. B. Schnabel. *Numerical Methods for Unconstrained Optimization and Nonlinear Equations*. SIAM, Philadelphia, 1996. Originally published by Prentice–Hall, 1983.

[DTT98]   T. A. Driscoll, K.-C. Toh, and L. N. Trefethen. From potential theory to matrix iterations in six steps. *SIAM Rev.*, 40:547–78, 1998.

[DV98]   T. A. Driscoll and S. A. Vavasis. Numerical conformal mapping using cross-ratios and Delaunay triangulation. *SIAM J. Sci. Comput.*, 19:1783–803, 1998.

[EH96]   A. R. Elcrat and C. Hu. Determination of surface and interior cracks from electrostatic measurements using Schwarz–Christoffel transformations. *Int. J. Eng. Sci.*, 34:1165–81, 1996.

[EIN95]   A. R. Elcrat, V. Isakov, and O. Neculoiu. On finding a surface crack from boundary measurements. *Inverse Problems*, 11:343–51, 1995.

[Elc82]   A. R. Elcrat. Separated flow past a plate with spoiler. *SIAM J. Math. Anal.*, 13:632–9, 1982.

[Ell83]   S. W. Ellacott. Computation of Faber series with application to numerical polynomial approximation in the complex plane. *Math. Comp.*, 40:575–87, 1983.

[ET86]   A. R. Elcrat and L. N. Trefethen. Classical free-streamline flow over a polygonal obstacle. *J. Comp. Appl. Math.*, 14:251–65, 1986.

[ET99]   M. Embree and L. N. Trefethen. Green's functions for multiply connected domains via conformal mapping. *SIAM Rev.*, 41:745–61, 1999.

[FA74]   K. Foster and R. Anderson. Transmission-line properties by conformal mapping. *Proc. Inst. Elec. Eng.*, 121:337–9, 1974.

[Fil61]   P. F. Filchakov. A method of determining the constants in the Schwarz–Christoffel integral. *Soviet Math. J.*, 2:877–81, 1961.

[Flo85]   J. M. Floryan. Conformal-mapping-based coordinate generation method for channel flows. *J. Comput. Phys.*, 58:229–45, 1985.

[Flo86]   J. M. Floryan. Conformal-mapping-based coordinate generation method for flows in periodic configurations. *J. Comput. Phys.*, 62:221–47, 1986.

[FPS99]   M. I. Falcão, N. Papamichael, and N. S. Stylianopoulos. Curvilinear crosscuts of subdivision for a domain decomposition method in numerical conformal mapping. *J. Comput. Appl. Math.*, 106:177–96, 1999.

[FZ87]   J. M. Floryan and C. Zemach. Schwarz–Christoffel mappings: A general approach. *J. Comput. Phys.*, 72:347–71, 1987.

[FZ88]   J. M. Floryan and C. Zemach. Quadrature rules for singular integrals with application to Schwarz–Christoffel mappings. *J. Comput. Phys.*, 75:15–30, 1988.

[FZ93]   J. M. Floryan and C. Zemach. Schwarz–Christoffel methods for conformal mappings of regions with periodic boundary. *J. Comp. Appl. Math.*, 46:77–102, 1993.

[Gai64]   D. Gaier. *Konstruktive Methoden der Konformen Abbildung*. Springer, Berlin, 1964.

[Gai72]   D. Gaier. Ermittlung des konformen Moduls von Vierecken mit Differenzenmethoden. *Numer. Math.*, 19:179–94, 1972.

[Gai87]   D. Gaier. *Lectures on Complex Approximation*. Birkhäuser, Boston, 1987. Translation of *Approximation im Komplexen*.

[GB⁺01]   M. Goano, F. Bertazzi, et al. A general conformal-mapping approach to the optimum electrode design of coplanar waveguides with arbitrary cross-section. *IEEE Microw. Theory Tech.*, 49:1573–80, 2001.

[GC87]   A. F. Ghoniem and Y. Cagnon. Vortex simulation of laminar recirculating flow. *J. Comput. Phys.*, 68:346–77, 1987.

[Gil49]   D. Gilbarg. A generalization of the Schwarz–Christoffel transformation. *Proc. Nat. Acad. Sci.*, 35:609–12, 1949.

[Gil60]   D. Gilbarg. Jets and cavities. In *Handbuch der Physik*, volume 9, pp. 311–445. Springer, Berlin, 1960.

[Goo50]   A. W. Goodman. On the Schwarz–Christoffel transformation and p-valent functions. *Trans. Amer. Math. Soc.*, 68:204–23, 1950.

[Goo60]   A. W. Goodman. Conformal mapping onto certain curvilinear polygons. *Univ. Nac. Tucumán Rev. Ser. A*, 13:20–6, 1960.

[GR88]   L. Greengard and V. Rokhlin. A fast adaptive multipole algorithm for particle simulations. *SIAM J. Sci. Stat. Comput.*, 9:669–86, 1988.

[GR94]   I. S. Gradshteyn and I. M. Ryzhik. *Table of Integrals, Series, and Products*, 5th edition. Academic Press, Boston, 1994.

[Gre96]   A. Greenbaum. Krylov subspace approximations to the solution of a linear system. In L. Adams and J. L. Nazareth, editors, *Linear and Nonlinear Conjugate Gradient-Related Methods*. SIAM, Philadelphia, 1996.

[Gui50]   E. A. Guillemin. *The Mathematics of Circuit Analysis*. Wiley, New York, 1950.

[Gur65]   M. I. Gurevich. *Theory of Jets in Ideal Fluids*. Academic Press, New York, 1965.

[GW69]   G. H. Golub and J. H. Welsch. Calculation of Gauss quadrature rules. *Math. Comp.*, 23:221–30, 1969.

[GZ94]   V. Y. Gutlyanskii and A. O. Zaidan. Contribution to the Kufarev method on determining the Schwarz–Christoffel parameters. *Dokl. Akad. Nauk*, 336:14–6, 1994.

[Hal67]   P. M. Hall. Resistance calculations for thin film patterns. *Thin Solid Films*, 1:277–95, 1967.

[Har78]   J. M. Harrison. The diffusion approximation for tandem queues in heavy traffic. *Adv. Appl. Prob.*, 10:886–905, 1978.

[HD86]   M. G. Harbour and J. M. Drake. Numerical method based on conformal transformations for calculating resistances in integrated circuits. *Internat. J. Electron.*, 60:679–89, 1986.

[Hel68]   H. Helmholtz. On discontinuous movements of fluids. *Phil. Mag.*, 36:337–46, 1868.

[Hen48]   P. Henrici. *Potentialprobleme mit scharfen und abgerundeten Ecken*. Master's thesis, Swiss Federal Inst. of Tech., Zurich, 1948.

[Hen74]   P. Henrici. *Applied and Computational Complex Analysis, Volume 1: Power Series, Integration, Conformal Mapping, Location of Zeros*. Wiley, New York, 1974.

[Hen86]   P. Henrici. *Applied and Computational Complex Analysis, Volume 3: Discrete Fourier Analysis, Cauchy Integrals, Construction of Conformal Maps, Univalent Functions*. Wiley, New York, 1986.

[Her82]   J. Hersch. Représentation conforme et symétries: Une détermination élémentaire du module d'un quadrilatère en forme de L. *Elem. der. Math.*, 37:1–5, 1982.

[Hil59]   E. Hille. *Analytic Function Theory*. Ginn, Boston, 1959.

[HKH94]   Y. Huang, D. J. Kouri, and D. K. Hoffman. General, energy-separable Faber polynomial representation of operator functions: Theory and application in quantum scattering. *J. Chem. Phys.*, 101:10493–506, 1994.

[HLD99]   M. Hassner, D. V. Leykin, and T. A. Driscoll. An analytic model of MR/GMR head sensitivity function. Technical Report RJ 10167 (95042), IBM Research Division, 1999.

[HLS85]   J. M. Harrison, H. J. Landau, and L. A. Shepp. The stationary distribution of reflected Brownian motion in a planar region. *Ann. Prob.*, 13:744–57, 1985.

[Hoe86]   M. Hoekstra. Coordinate generation in symmetrical interior, exterior, or annular 2D domains, using a generalized Schwarz–Christoffel transformation. In J. Hauser and C. Taylor, editors, *Numerical Grid Generation in Computational Fluid Mechanics*. Pineridge Press, Swansea, UK, 1986.

[Hou89]   D. M. Hough. Conformal mapping and Fourier–Jacobi approximations. ETH Zürich IPS Research Report 89-06, 1989.

[Hou90]   D. M. Hough. User's guide to CONFPACK. ETH Zürich IPS Research Report 90-11, 1990.

[How73]   D. Howe. The application of numerical methods to the conformal transformation of polygonal boundaries. *J. Inst. Math. Applic.*, 12:125–36, 1973.

[How90]   L. H. Howell. *Computation of Conformal Maps by Modified Schwarz–Christoffel Transformations.* PhD thesis, MIT, 1990.

[How93]   L. H. Howell. Numerical conformal mapping of circular arc polygons. *J. Comp. Appl. Math.*, 46:7–28, 1993.

[How94]   L. H. Howell. Schwarz–Christoffel methods for multiply-elongated regions. In *Proc. of the 14th IMACS World Congress on Computation and Applied Mathematics*, Atlanta, 1994.

[HP78]   E. Haugeneder and W. Prochazka. Automatische Berechnung der Durchbiegungen und der Schnittgrössen dünner Platten mit Hilfe der Funktionentheorie. *Bauingenieur*, 53:243–8, 1978.

[HP83]   D. M. Hough and N. Papamichael. An integral equation method for the numerical conformal mapping of interior, exterior, and doubly-connected domains. *Numer. Math.*, 41:287–307, 1983.

[HR79]   T. R. Hopkins and D. E. Roberts. Kufarev's method for determining the Schwarz–Christoffel parameters. *Numer. Math.*, 33:353–65, 1979.

[HT90]   L. H. Howell and L. N. Trefethen. A modified Schwarz–Christoffel transformation for elongated regions. *SIAM J. Sci. Stat. Comput.*, 11:928–49, 1990.

[Hu95]   C. Hu. User's guide to DSCPACK. Nat. Inst. Aviation Res. 95-1, Wichita State Univ., 1995.

[Hu98]   C. Hu. Algorithm 785: A software package for computing Schwarz–Christoffel conformal transformation for doubly connected polygonal regions. *ACM Trans. Math. Soft.*, 24(3):317–33, 1998.

[Hug75]   O. F. Hughes. A simplification of the Schwarz–Christoffel formula for symmetric quadrilateral transformation. *SIAM J. Numer. Anal.*, 6:258–61, 1975.

[Ive82]   D. C. Ives. Conformal grid generation. *Appl. Math. Comput.*, 10/11:107–35, 1982. Appears in the collection *Numerical Grid Generation*, J. F. Thompson, editor, pp. 107–35. North–Holland, New York, 1982.

[JM87]   K. P. Jackson and J. C. Mason. The approximation by complex functions of stresses in cracked domains. In J. C. Mason and M. G. Cox, editors, *Algorithms for Approximation*, pp. 611–22. Oxford University Press, Oxford, 1987.

[Joh82]   F. John. *Partial Differential Equations,* 4th edition. Springer, New York, 1982.

[Kir69]   G. Kirchhoff. Zur Theorie freier Flüssigkeitsstrahlen. *J. Reine Angew. Math.*, 70:289–98, 1869.

[KK64]   L. V. Kantorovich and V. I. Krylov. *Approximate Methods of Higher Analysis,* 3rd edition. Interscience, New York, 1964.

[KO89]   Ç. K. Koç and P. F. Ordung. Schwarz–Christoffel transformation for the simluation of two-dimensional capacitance. *IEEE Trans. Comp.-Aided Des.*, 8:1025–7, 1989.

[Kom45]   Y. Komatu. Darstellungen der in einem Kreisringe analytischen Funktionen nebst den Anwendungen auf konforme Abbildung über Polygonalringgebiete. *Jap. J. Math.*, 19:203–15, 1945.

[Kos94]   R. Kosloff. Propagation methods for quantum molecular dynamics. *Annu. Rev. Phys. Chem.*, 45:145–78, 1994.

[KR88]   B. C. Krikeles and R. L. Rubin. On the crowding of parameters associated with Schwarz–Christoffel transformations. *Appl. Math. Comp.*, 28:297–308, 1988.

[Küh83]   R. Kühnau. Numerische Realisierung konformer Abbildungen durch "Interpolation". *Z. Angew. Math. Mech.*, 63:631–7, 1983.

[Kyt98]   P. K. Kythe. *Computational Conformal Mapping*. Birkhäuser, Boston, 1998.

[Lam45]   H. Lamb. *Hydrodynamics*. Dover, New York, 1945.

[Lan87]   S. Lang. *Elliptic Functions,* 2nd edition. Springer-Verlag, New York, 1987.

[Lau94]   R. Laugesen. Conformal mapping of long quadrilaterals and thick doubly connected domains. *Constr. Approx.*, 10:523–54, 1994.

[Lea15]   J. G. Leathem. Some applications of conformal transformation to problems in hydrodynamics. *Phil. Trans. Roy. Soc. Lond.*, 215:439–87, 1915.

[LG68]   P. J. Lawrenson and S. K. Gupta. Conformal transformation employing direct-search techniques of minimisation. *Proc. IEE*, 115:427–31, 1968.

[Li92]   X. Z. Li. An adaptive method for solving nonsymmetric linear-systems involving applications of SCPACK. *J. Comput. Appl. Math.*, 44:351–70, 1992.

[Lis99]   V. D. Liseikin. *Grid Generation Methods*. Springer, Berlin, 1999.

[LL83]   P. D. Lax and C. D. Levermore. The small dispersion limit of the Korteweg–de Vries equation II. *Comm. Pure Appl. Math.*, 36:571–93, 1983.

[Mey79]   E.-S. Meyer. *Praktische Verfahren zur konformen Abbildung von Geradenpolygonen*. PhD thesis, Universität Hannover, 1979.

[Mil96]   K. G. Miller. Stationary corner vortex configurations. *Z. Angew. Math. Phys.*, 47:39–56, 1996.

[Mon83]   V. N. Monakhov. *Boundary-Value Problems with Free Boundaries for Elliptic Systems of Equations*. Amer. Math. Soc., Providence, RI, 1983.

[MRB94]   K. B. McGrattan, R. G. Rehm, and H. R. Baum. Fire-driven flows in enclosures. *J. Comput. Phys.*, 110:285–91, 1994.

[MZ80]   R. Menikoff and C. Zemach. Methods for numerical conformal mapping. *J. Comput. Phys.*, 36:366–410, 1980.

[Nee97]   T. Needham. *Visual Complex Analysis*. Clarendon Press, Oxford, 1997.

[Neh52]   Z. Nehari. *Conformal Mapping*. Dover, New York, 1952.

[Nev93]   O. Nevanlinna. *Convergence of Iterations for Linear Equations*. Birkhäuser, Basel, 1993.

[New79]   G. F. Newell. *Approximate Behavior of Tandem Queues*, volume 171 of *Lecture Notes in Econ. and Math Systems*. Springer, Berlin, 1979.

[Nic77]   A. Nicolaide. Numerical methods in conformal transformation. *Proc. Inst. Elec. Eng. Lon.*, 124:1110, 1977.

[Nic97]   O. Nicolio. Numerical computations on the m-resistor trimming problem. *Comput. Ind. Eng.*, 33:393–6, 1997.

[OL96]   F. Olyslager and I. V. Lindell. Capacitance relations for a class of two-dimensional conductor configurations. *IEE Proc. Sci. Meas. Technol.*, 143:302–8, 1996.

[Pal37]   H. B. Palmer. The capacitance of a parallel-plate capacitor by the Schwarz–Christoffel transformation. *Trans. Amer. Inst. Elec. Engr.*, 56:363–6, 1937.

[Pea91]   K. Pearce. A constructive method for numerically computing conformal
           mappings for gearlike domains. *SIAM J. Sci. Stat. Comput.*, 12:231–46,
           1991.

[PH95]   M. C. A. M. Peters and H. W. M. Hoeijmakers. A vortex sheet method
           applied to unsteady flow separation from sharp edges. *J. Comp. Phys.*,
           120:88–104, 1995.

[PM63]   V. L. Pisacane and L. E. Malvern. Application of numerical mapping to
           the Muskhelishvili method in plane elasticity. *J. Appl. Mech.*, 30:410–14,
           1963.

[PM72]   T. W. Parks and J. H. McClellan. Chebyshev approximation for nonrecursive
           digital filters with linear phase. *IEEE Trans. Circuit Theory*, CT-19:189–94,
           1972.

[Pro83]   W. Prochazka. Conformal mapping of the unit circle or of the upper half
           plane onto a polygon. *Computing*, 31:155–72, 1983.

[PS91]   N. Papamichael and N. S. Stylianopoulos. A domain decomposition method
           for conformal mapping onto a rectangle. *Constr. Approx.*, 7:349–79,
           1991.

[PS99]   N. Papamichael and N. S. Stylianopoulos. The asymptotic behavior of con-
           formal modules of quadrilaterals with applications to the estimation of
           resistance values. *Contr. Approx.*, 15:109–34, 1999.

[Rep79]   K. Reppe. Berechnung von Magnetfeldern mit Hilfe der konformen Abbil-
           dung durch numerische Integration der Abbildungsfunktion von Schwarz–
           Christoffel. *Siemens Forsch. u. Entwickl. Ber.*, 8:190–5, 1979.

[RS87]   B. Rodin and D. Sullivan. The convergence of circle packings to the
           Riemann mapping. *J. Differential Geom.*, 26:349–60, 1987.

[Sch69a]   H. A. Schwarz. Conforme Abbildung der Oberfläche eines Tetraeders auf
           die Oberfläche einer Kugel. *J. Reine Ange. Math.*, 70:121–36, 1869. Also
           in the collected works [Sch90], pp. 84–101.

[Sch69b]   H. A. Schwarz. Über einige Abbildungsaufgaben. *J. Reine Ange. Math.*,
           70:105–20, 1869. Also in the collected works [Sch90], pp. 65–83.

[Sch90]   H. A. Schwarz. *Gesammelte Mathematische Abhandlungen*, volume II.
           Springer, Berlin, 1890.

[Sch98]   C. Schwab. *p- and hp-Finite Element Methods. Theory and Applications
           in Solid and Fluid Mechanics*. Oxford University Press, New York, 1998.

[SD85]   K. P. Sridhar and R. T. Davis. A Schwarz–Christoffel method for generating
           two-dimensional flow grids. *J. Fluids Eng.*, 107:330–7, 1985.

[Sku66]   R. S. Skulsky. A conformal mapping method to predict low-speed aerody-
           namic characteristics of arbitrary slender re-entry shapes. *AIAA J. Space-
           craft*, 3:247–53, 1966.

[SL91]   R. Schinzinger and P. A. A. Laura. *Conformal Mapping: Methods and
           Applications*. Elsevier, Amsterdam, 1991.

[Sno99]   J. Snoeyink. Cross-ratios and angles determine a polygon. *Discrete Comput.
           Geom.*, 22:619–31, 1999.

[Squ75]   W. Squire. Computer implementation of Schwarz–Christoffel transforma-
           tion. *J. Franklin Inst. Eng. Appl. Math.*, 299:315–21, 1975.

[SS99]   J. Shen and G. Strang. The asymptotics of optimal (equiripple) filters. *IEEE
           Trans. Signal Proc.*, 47:1087–98, 1999.

[SSW01] J. Shen, G. Strang, and A. J. Wathen. The potential theory of several intervals and its applications. *Appl. Math. Optim.*, 44:67–85, 2001.

[ST84] P. G. Saffman and S. Tanveer. Vortex induced lift on two-dimensional low speed wings. *Stud. Appl. Math.*, 71(1):65–78, 1984.

[Sta93] G. Starke. Fejér–Walsh points for rational functions and their use in the ADI iterative method. *J. Comput. Appl. Math.*, 46:129–41, 1993.

[Ste99] K. Stephenson. Approximation of conformal structures via circle packing. In N. Papamichael, S. Ruscheweyh, and E. B. Saff, editors, *Computational Methods and Function Theory 1997*, pp. 551–82. World Scientific, Singapore, 1999.

[SV93] G. Starke and R. S. Varga. A hybrid Arnoldi–Faber iterative method for nonsymmetric systems of linear equations. *Numer. Math.*, 64:213–40, 1993.

[TD98] L. N. Trefethen and T. A. Driscoll. Schwarz-Christoffel mapping in the computer era. In *Proceedings of the International Congress of Mathematicians, Vol. III (Berlin, 1998)*, volume 1998, pp. 533–42 (electronic), 1998.

[Toz83] O. V. Tozoni. Computer-simulation of conformal-mappings. *Cybernetics*, 19:464–74, 1983.

[Tre80] L. N. Trefethen. Numerical computation of the Schwarz–Christoffel transformation. *SIAM J. Sci. Stat. Comput.*, 1:82–102, 1980.

[Tre84] L. N. Trefethen. Analysis and design of polygonal resistors by conformal mapping. *Z. Angew. Math. Phys.*, 35:692–704, 1984.

[Tre86] L. N. Trefethen, editor. *Numerical Conformal Mapping*. North-Holland, Amsterdam, 1986. Reprint of *J. Comput. Appl. Math.*, 14, 1986.

[Tre89] L. N. Trefethen. SCPACK user's guide. MIT Numerical Analysis Report 89-2, 1989.

[Tre93] L. N. Trefethen. Schwarz–Christoffel mapping in the 1980's. Cornell University Computer Science Department Technical Report TR 93-1381, 1993.

[TSW99] J. F. Thompson, B. K. Soni, and N. P. Weatherill. *Handbook of Grid Generation*. CRC Press, Boca Raton, FL, 1999.

[TW86] L. N. Trefethen and R. J. Williams. Conformal mapping solution of Laplace's equation on a polygon with oblique derivative boundary conditions. *J. Comp. Appl. Math.*, 14:227–49, 1986.

[Ver82] W. Versnel. The geometric correction factor for a rectangular Hall plate. *J. Appl. Phys.*, 53:4980–6, 1982.

[Ver83] W. Versnel. Electrical characteristics of an anisotropic semiconductor sample of circular shape with finite contacts. *J. Appl. Phys.*, 54:916–21, 1983.

[VK74] V. V. Vecheslavov and V. I. Kokoulin. Determination of the parameters of the conformal mapping of simply connected polygonal regions. *USSR Comput. Math. Math. Phys.*, 13:57–65, 1974.

[Vol79] E. A. Volkov. Exponentially converging method of the conformal mapping of multi-angular domains. *Dok. Akad. Nauk SSSR*, 249:1292–5, 1979.

[Vol87] E. A. Volkov. Approximate conformal mapping of certain polygons onto a strip by the block method. *USSR Comp. Math. Math. Phys.*, 27:136–42, 1987.

[Vol88] E. A. Volkov. Approximate conformal mapping by the block method of a circle with a polygonal hole into a ring. *USSR Comp. Math. Math. Phys.*, 28:143–7, 1988.

[vS59]   W. von Koppenfels and R. W. Stallman. *Praxis der Konformen Abbildung.* Springer, Berlin, 1959.

[VT73]   V. V. Vecheslavov and N. I. Tolstobrova. Parameters for the conformal mapping of symmetric doubly connected polygonal regions (English translation). *USSR Comput. Math. Math. Phys.*, 13:49–57, 1973.

[Wal64]   M. Walker. *The Schwarz–Christoffel Transformation and Its Applications— A Simple Exposition.* Dover, New York, 1964. First published in 1933 by Oxford University under the title, *Conjugate Functions for Engineers.*

[Weg86]   R. Wegmann. An iterative method for conformal mapping. *J. Comp. Appl. Math.*, 14:7–18, 1986.

[Wen82]   M. Wenocur. *A Production Network Model and its Diffusion Approximation.* PhD thesis, Department of Statistics, Stanford University, 1982.

[Wic54]   R. F. Wick. Solution of the field problem of the Germanium gyrator. *J. Appl. Phys.*, 25:741–56, 1954.

[Wid69]   H. Widom. Extremal polynomials associated with a system of curves in the complex plane. *Adv. Math.*, 3:127–232, 1969.

[Wil95]   R. J. Williams. Semimartingale reflecting Brownian motions in the orthant. In *Stochastic Networks*, pp. 125–37. Springer, New York, 1995.

[Woo61]   L. C. Woods. *The Theory of Subsonic Plane Flow.* Cambridge University Press, New York, 1961.

Index

131

Printed in the United States
By Bookmasters